REIMAGINING SAMPLE-BASED HIP HOP

Reimagining Sample-based Hip Hop: Making Records within Records presents the poetics of hip-hop record production and the significance of sample material in record making, providing analysis of key releases in hip-hop discography and interviews with experts from the world of Hip Hop and beyond.

Beginning with the history of hip-hop music making, this book guides the reader through the alternative techniques deployed by beat-makers to avoid the use of copyrighted samples and concludes with a consideration of the future of Hip Hop, alongside a companion album that has been created using findings from this research. Challenging previous theoretical understandings about Hip Hop, the author focuses on deconstructing sonic phenomena using his hands-on engineering expertise and in-depth musicological knowledge about record production.

With a significant emphasis on both practice and theory, *Reimagining Sample-based Hip Hop* will be of interest to advanced undergraduates, postgraduates, and researchers working in audio engineering, music production, hip-hop studies, and musicology.

Michail Exarchos (aka Stereo Mike) is a hip-hop musicologist and award-winning rap artist (first MTV Best Greek Act winner), with nominations for seven video music awards and an MTV Europe Music Award. He is the creative director of music innovation company RT60 Ltd., specialising in intelligent apps, and has led programmes in recording, mixing, mastering, and production at various UK institutions. Under his Stereo Mike alias, he has produced three critically acclaimed solo albums, and numerous singles for international artists on labels such as EMI, Sony, Universal, and Warner Music. Mike's self-produced album, *XLi3h*, has been included in the 30 Best Greek hip-hop albums of all time.

"This book needed to be written. We're lucky the job was done by such a creative soul, expert engineer, and insightful scholar. The gift we're left with is a penetrating exposition of the poetic fabric of contemporary music recording practice and culture."

Albin Zak, *Professor of Music Emeritus, University at Albany*

"*Reimagining Sample-based Hip Hop* is a tour de force and the best example I have seen of theory and practice coming together in hip-hop studies to date. Not only does this book provide a useful template for practice-as-research based methods, but it is rich in theorizing sampling, remix and beatmaking cultures. Exarchos has combined a number of worlds in ways that we were all waiting for someone to do expertly."

Justin A. Williams, *Associate Professor of Music, University of Bristol*

"*Reimagining Sample-based Hip Hop* provides deep insight into the rich tapestry of underlying layers that embody sample-based Hip-Hop. Michail Exarchos meticulously and clearly poses alternative practices to Hip-Hop production and to beat-making, using originally constructed source material rather than previously released phonographic content. The reader will benefit from the strong coverage of authenticity and mechanical borrowing as applied to a great breadth of music and production approaches that define the style's unique sonic signature of material 'play' with phonographic objects. This is a must read for anyone interested in the workings of or creation of Hip-Hop."

William Moylan, *Professor of Music and Sound Recording Technology, University of Massachusetts Lowell*

Perspectives on Music Production

Series Editors: Russ Hepworth-Sawyer, *York St John University, UK,* **Jay Hodgson**, *Western University, Ontario, Canada, and* **Mark Marrington**, *York St John University, UK*

This series collects detailed and experientially informed considerations of record production from a multitude of perspectives, by authors working in a wide array of academic, creative and professional contexts. We solicit the perspectives of scholars of every disciplinary stripe, alongside recordists and recording musicians themselves, to provide a fully comprehensive analytic point-of-view on each component stage of music production. Each volume in the series thus focuses directly on a distinct stage of music production, from pre-production through recording (audio engineering), mixing, mastering, to marketing and promotions.

Recording the Classical Guitar
Mark Marrington

The Creative Electronic Music Producer
Thomas Brett

3-D Audio
Edited by Justin Paterson and Hyunkook Lee

Understanding Game Scoring
The Evolution of Compositional Practice for and through Gaming
Mackenzie Enns

Coproduction
Collaboration in Music Production
Robert Wilsmore and Christopher Johnson

Distortion in Music Production
The Soul of Sonics
Edited by Gary Bromham and Austin Moore

Reimagining Sample-based Hip Hop
Making Records within Records
Michail Exarchos

For more information about this series, please visit: www.routledge.com/Perspectives-on-Music-Production/book-series/POMP

REIMAGINING SAMPLE-BASED HIP HOP

Making Records within Records

Michail Exarchos

LONDON AND NEW YORK

Designed cover image: Michail Exarchos

First published 2024
by Routledge
4 Park Square, Milton Park, Abingdon, Oxon OX14 4RN

and by Routledge
605 Third Avenue, New York, NY 10158

Routledge is an imprint of the Taylor & Francis Group, an informa business

British Library Cataloguing-in-Publication Data
A catalogue record for this book is available from the British Library

Library of Congress Cataloging-in-Publication Data
Names: Exarchos, Michail, author.
Title: Reimagining sample-based hip hop : making records within records / Michail Exarchos.
Description: Abingdon, Oxon ; New York : Routledge, 2024. |
Series: Perspectives on music production | Includes bibliographical references and index. |
Identifiers: LCCN 2023004128 (print) | LCCN 2023004129 (ebook) |
ISBN 9780367461805 (paperback) | ISBN 9780367461812 (hardback) |
ISBN 9781003027430 (ebook)
Subjects: LCSH: Rap (Music)—History and criticism. | Sampler (Musical instrument)—History. | Rap (Music)—Production and direction—History.
Classification: LCC ML3531 .E98 2024 (print) | LCC ML3531 (ebook) |
DDC 782.421649—dc23/eng/20230414
LC record available at https://lccn.loc.gov/2023004128
LC ebook record available at https://lccn.loc.gov/2023004129

ISBN: 978-0-367-46181-2 (hbk)
ISBN: 978-0-367-46180-5 (pbk)
ISBN: 978-1-003-02743-0 (ebk)

DOI: 10.4324/9781003027430

Typeset in Bembo
by codeMantra

Access the Support Material: www.stereo-mike.com

For Lili and Dino

CONTENTS

FIGURES

TABLES

PREFACE

Reimagining Sample-based Hip Hop: Making Records within Records deals with the poetics of hip-hop record production making use of originally constructed source material, rather than previously released phonographic content.[1] The idea behind the project was borne out of a practical conundrum I faced when, in the mid-2000s, I gradually transitioned from underground beat-maker to a major-label tenure with EMI Music (Greece) as a signed artist/producer.[2] The leap towards a mainstream/national profile highlighted acute issues with sample-based music making, and brought me face-to-face with a larger phenomenon characterising world-wide beat-making practice. There is a tangible gap that exists between underground and mainstream sample-based hip-hop paradigms, dictated by an increasingly restrictive licensing landscape, and for the majority of producers operating between the two extremes, access to raw phonographic sources has become a pursuit of outermost importance. This pursuit has driven beat-makers to seek innovative alternatives to practising the sample-based artform, whilst avoiding the licensing and financial ramifications tied to copyrighted sample use. This book investigates the growing practice of sample-*creating*-based Hip Hop, through both theoretical analysis and creative practice.

Inspired by a parallel career in academia spanning two decades as a music technology and audio production educator, the examination of the phenomenon culminated in a practice-based doctoral research project grounded in the musicology of record production; and so this book has its roots in a PhD, albeit drawing from my industry experience as a hip-hop artist, audio engineer, and music producer. As such, it is accompanied by a previously unreleased large-scale instrumental album of original beats, supplying an applied investigative context, and forming a *soundtrack* unfolding in parallel to the text. The largely autoethnographic lens I deploy over the creative practice attempts to relate the *personal* in my beat-making approach to the larger cultural/aesthetic phenomenon, and the research is further supported by interviews with expert practitioners, as well as phonographic/aural and literary analysis. The theoretical and practical findings drawn from the investigation illuminate an unexamined practice with profound impact upon popular music culture.

This book's research narrative is therefore capable of reshaping our understanding of creative beat-making in the flux of a shifting legal and pragmatic landscape. Furthermore, the non-linear, juxtaposed, and arguably metamodern dimensions of the practice readdress current/historical debates about Hip Hop, putting *sonic materiality* at the forefront of the discussion, and challenging the methodological strategies deployed thus far for the study of contemporary, electronic, and Afrological music forms. Consequently, an identifiable need arises for a thorough exploration of—and theorising upon—this form of record production practice. From Dr. Dre, through to De La Soul, J.U.S.T.I.C.E. League, Boards of Canada, Portishead, Statik Selektah, Marco Polo, Griselda Records associates, and Frank Dukes, sample-based hip-hop producers have creatively renegotiated the landscape surrounding sample deployment through alternative production approaches. These techniques rely on the creation of interim sampling content for subsequent use in what can be described as a form of 'meta' phonographic process: an innovative phenomenon with important creative implications powering some of today's biggest hits; and—arguably—an evolutionary strategy facilitating the future development of the genre (and sample-based music as a whole). In the aesthetic pursuit of what makes a newly created source sufficiently 'phonographic' in the context of sample-based Hip Hop, this text (re)addresses the way in which we consider how the sonic past interacts with the music present, and extrapolates upon the way in which such a musical practice may mirror a metamodern zeitgeist in other arts, and culture as a whole.

Notes

1 For consistency, Hip Hop will appear capitalised when used as a noun and referring to the musical genre, and hyphenated, in lowercase, when used as an adjective to describe ensuing nouns (for instance, process, artist, or production). Many of the cited authors in this book opt for Hip-Hop with a hyphen (in lowercase or capitalised), and their chosen conventions will be respected within quotations. This book's convention, however, will extend to other genres and styles.

2 The terms 'beat-maker/making' will be used interchangeably with 'sample-based hip-hop producer/production'. 'Beat', in hip-hop parlance, refers to a complete instrumental music production or backing, not just the organisation of percussive/drum elements, which highlights the genre's rhythmic priorities. Williams (2010, p. 19) extends Schloss's (2004/2014, p. 2) definition of 'beat' as a sample-based instrumental collage "composed of brief segments of recorded sound", to also include non-sample-based elements in the instrumental production.

Bibliodiscography

Schloss, J.G. (2014) *Making beats: The art of sample-based Hip-Hop*. Middletown, CT: Wesleyan University Press (Music/Culture).

Williams, J.A. (2010) *Musical borrowing in hip-hop music: Theoretical frameworks and case studies*. Unpublished PhD thesis. University of Nottingham.

ACKNOWLEDGEMENTS

This book and the associated album of beats represent the culmination of a life-long journey deeply listening to, making, teaching, learning from, and analysing Hip Hop. The book itself has evolved out of a PhD thesis, while the accompanying album doubles up as an independent release and sonic manifestation of the phonographic phenomena deciphered in the text.

I am grateful to my doctoral supervisors Professors Justin Paterson and Simon Zagorski-Thomas for the support, encouragement, advice, and warm welcome to the inspiring research community at London College of Music (LCM). Specifically, I want to thank Professor Justin Paterson for the attention to detail he has extended across all milestones of the research journey that led to this book, and our late-night inter-stylistic discussions on rhythm and timbre ranging from India, to Africa, Scotland, and the Balkans; also, for the wonderful sonic collaborations, whether remixing Hip Hop for new interactive platforms, or multi-miking cathedral organs at Arundel and Coventry; as well as the tireless and educational behind-the-scenes collaborative efforts in organising conferences, such as London's *Innovation in Music 2019*. I want to thank Professor Simon Zagorski-Thomas for inviting me into the LCM family—a result of meeting at his *Art of Record Production* conferences, which are in turn responsible for creating a vibrant network of enthusiastic phono-musicologists. Simon's pertinent questions about the work discussed in this book led to essential 'eureka' moments, particularly with regard to sonic materiality, the 'phonographic' essence of sampled instances, and the metamodern dimensions of the practice.

My thanks also extend to Dr Tim Hughes, who advised at important milestones of the journey, and whose knowledge of African-American music has been invaluable to the shape and depth this book has assumed. From Stevie Wonder, to Hendrix, harpsichords, and the MPC, Tim helped me make associations between my evolving praxis, and Afrological musical priorities deeply embedded in the tools and utterances of a worldwide hip-hop syntax. Professor Jonathan Stockdale was instrumental in helping me set the methodological foundations behind this book's research design. And Dr Andrew Bourbon provided sonic inspiration via our spatial hip-hop performances around the world and the endless pursuit of ideal mix 'staging'.

The portfolio of beats produced as part of the album accompanying this book is a result of harmonious collaboration with a large musical family. Professor Rob Toulson and Dr Paul Thompson are my musical partners in band FET 47, which features in a number of samples providing inspiration and sonic content for the album's beats. Rob also opened up the door for my attendance and presentation at numerous international conferences with his concrete support during our tenure at the University of Westminster. Paul Thompson's live drum performances provide the backbone for a plethora of beats (Paul, you are my 'funky drummer'). Jo Lord is my musical partner in duet collaborations ranging from Americana to Grunge, Punk, and dark Electronica, and her voice punctuates the hours of boom-bap rhythms with precious melodies and lyricism. Andy Caldecott's punk slam poetry brings much-needed vocal darkness, political commentary, and aggression in key moments. Dr Sara McGuinness welcomed me as an engineer (and MC) to London's Latin and African communities, specifically via our production collaboration on bands Sarabanda and Grupo Lokito. As a result, the beats are enriched with a universal tapestry of sounds and performances, such as percussion (by Bill Bland, Tristan Butler, and David Pattman), voices (by Elpidio Caicedo Alegria, Eugene Makuta, and Emeris Solis), horns (by Clare Hirst, Viva Msimang, and Deanna Wilhelm), guitar (Kiamfu 'Burkina Faso' Kasongo), and bass (Elpidio Caicedo Alegria). Our old cassette recordings of free-jazz experiments with childhood friend and collaborator Antonios Tsoukatos (on electric guitar) became a new 'instrument' under pedal and variplayed Walkman manipulation; our indie/folk jams with Dr Dan Pratt fused Greek baglama explorations with Doors-inspired keyboard improvisations; and Albin Zak gracefully allowed me to reimagine his productions, 'Your Outer Atmosphere' and 'One Another'.

My thanks extend to doctoral examiners Dr Justin Williams and Dr Dave Hook, who have provided valuable feedback and insight towards the completion of this journey, through their critical examination of the text and practice, and an inspiring and helpful Viva experience. Professors William Moylan, Justin Williams, and Albin Zak have cast a critical eye (and ear) over the book (and album), and I am indebted to them for their kind attention, as well as their ground-breaking scholarship, without which this book would not have been possible. Russ Hepworth-Sawyer and Mark Marrington kindly invited me into the Perspectives on Music Production book series, providing thoughtful editorial support, while Russ lovingly mastered the large body of beats. Thank you Hannah Rowe and the editorial team at Routledge/Focal Press, too, for the thorough collaboration during this—dual—undertaking.

Last but not least, I am grateful to my parents (Lili and Dino) for supporting me throughout this life-long hip-hop journey, the last leg of which includes this book/beatscape adventure. Specifically, I want to thank my mother, Lili, for taking me—as a disillusioned adolescent—to that outreach interview in Athens, Greece, which brought me to UK academia (the 'Dr' in this whole effort goes to you).

INTRODUCTION

Since its inception, hip-hop music has been defined by the use, appropriation, and re-contextualisation of found sound. From Kool Herc's turntablism over funk records that helped invent the genre (circa 1973), through to the Beastie Boys' almost entirely sample-based *Paul's Boutique* (1989), and all the way to sample-heavy resurgences in contemporary hip-hop production, phonographic samples have provided the foundation, inspiration, and characteristic sonic footprint for large parts of the style's output. On the other hand, changes in sampling legislation and increasing sample-clearing costs have had a profound effect on hip-hop production practices, often resulting in the replacement, reduction, or complete omission of phonographic samples for notable periods of its trajectory.[1] These alternative production methods, however, have led to what has been described as a "softer" (Shocklee, 2004) or less authentic phonographic footprint (Marshall, 2006; Thompson and Greenman, 2013, p. 101), which highlights the *sonic* implications that a sample-free process impacts on the end artefacts. This realisation holds the potential to shift the practitioner's focus from Hip Hop's historical association with 'musical borrowing' to the significance of '*mechanical borrowing*'.[2]

Writing about Jay-Z's *The Blueprint* (2001) on the 20th anniversary of the album's release in a recent Instagram post, producer 9th Wonder (2021) accurately captures the interplay between poetics, the licensing landscape, and resulting aesthetics in Hip Hop's trajectory:

> SAMPLING IS DEAD.....
>
> Or at least that's what I was told during that time. Outside of @hitek [Hitek], [J] DILLA, Madlib, @realpeterock [Pete Rock], @djpremier [DJ Premier], and so many others still digging and chopping up records, the mainstream at large somewhat chose to abandon the notion.[3] Producers were becoming more in tune with their publishing rights, and, staying away from artists who took away a majority of their share. I chose to however, keep digging despite what the naysayers said.[4] On a mainstream level, it takes ONE album to turn things around, there may be precursors, but the one album that put soul back in mainstream hip-hop, was this one [*The Blueprint* (Jay-Z, 2001)]. @justblaze [Just Blaze], @kanyewest [Kanye West], and @binkthehumblemonsta

DOI: 10.4324/9781003027430-1

> [Bink] was the right cast of producers who gave Jay-Z the *perfect chunky sound beds* to bare his soul.
>
> *(my emphasis)*

Project hypothesis

The underlying premise in this book, thus, is that the appeal of the sample-based production aesthetic in Hip Hop is the result of a 'mechanical borrowing' that captures a multi-layered sonic imprint of historical recording, mixing, mastering, and even physical manufacturing processes. This, in turn, has implications not only for the end outputs but also for the creative and technical poetics that it inspires. Many hip-hop producers who favour sampling have expressed explicit links between the genre's perceived authenticity and the use of samples, particularly ones capturing past or analogue eras (Krims, 2000; Collins, 2008; Harkins, 2008; Schloss, 2014).[5] Existing literature attributes this perception to the technology deployed, the spaces captured, historical musicianship trends, and the creative workflow determined by the sampling process itself (Rose, 1994; Krims, 2000; Harkins, 2008; Sewell, 2013; Schloss, 2014).

Conversely, practitioners' opinions appear polarised on the use of live instrumentation in Hip Hop as an authentic production alternative (Marshall, 2006). The distinction that I will draw out throughout this book is that the majority of samples deployed in hip-hop practice *do* feature live instrumentation inherently (albeit previously recorded, produced, and released), but it is the nature of the production that contains them which further shapes the authenticity argument, and determines the sonics of the end artefacts. This delineation echoes Allan Moore's (2002, pp. 211–218) notions of "first" and "third person authenticity", acknowledging that sample-based aesthetics may rely, respectively, on the recognition of specific (artist) utterances or generic (category) signifiers.[6] But it problematises how the latter *perception* is constructed, by examining the mechanics that shape an adequate phonographic context (or in Moore's (2002, p. 216) words: "the fabrication of ... a conceptual (if not historical) point of origin"). The project is therefore a study in examining and creating phonographic—rather than cultural—history, whilst acknowledging the cultural factors that shape the *context* that provides meaning to the *content* (the material elements) of the sonic domain. The practical problem posed is whether there can be effective production alternatives towards achieving the desired aesthetic objectives, simultaneously avoiding the use of copyrighted samples. This project explores the hypothesis through a design that incorporates a large-scale (instrumental) album production process, referencing a range of classic musical epochs and styles at source level. The aim of the investigation is to create—and highlight the strategies responsible for producing—original hip-hop output from self-made record segments that communicate *convincing* phonographic qualities. In pursuit of the practice-based vision, a number of essential and associated objectives are also tackled:

- exploring vintage production techniques that can be deployed in pursuit of phonographic constructs effectively, which facilitate sample-based composition;
- reviewing literature on the history, musicology, and interaction of record production, sampling, and hip-hop aesthetics;
- assessing peers' opinions on sample-based aesthetics and processes in contemporary beat-making; and
- investigating the interaction between beat-making and the production of original source content.

Background

Previous literature on Hip Hop has overwhelmingly favoured notions of "musical borrowing", which—even when inclusive of broad themes such as the intertextuality and reception of genre, the rapper's voice, and sampled sound (see, for example: Williams, 2010)—promote elements of the musical and/or lyrical domains (musical style, motifs, words, and rhythm) as primary foci. On the occasions when sonic texture, material content, and the overall phonographic timbre resulting from sample-based processes are considered, the majority of studies stop at surface observations. For example, Tricia Rose's otherwise foundational mapping of Hip Hop's sonic priorities to "black intent" in *Black noise* (1994), mostly "focuses on the conventions of orality in hip-hop rather than on the underlying sonic product that supports and sustains it" (Goldberg, 2004, p. 130). Schloss's exhaustive ethnographic study of Golden Age practices in *Making beats* (2014)—which accurately identifies the sonic characteristics that define the period's sample-based aesthetic—only extends as far as *filtering* techniques with regards to the sonic domain (deployed in the service of sound-object isolation).[7] Analysis of essential processes such as (swing) quantisation and sample chopping revert back to rhythmic and structural (i.e. motivic) ramifications. Adam Krims's (2000, pp. 41–54) Kantian-inspired "hip-hop sublime" extends as far as recognising "timbre ... as a crucial means of organization" and a catalyst in the listener's sense-making sonic experience; but it does not explore how the "combination of incommensurable musical layers ... are selectively and dramatically brought into conflict with each other". Krims's timbral sublime does echo William Moylan's (2020, p. 239) "gestalt percept" as a "singular impression that coalesces from its many parts"; but Moylan goes further by considering the acoustic component parts that make up this *timbral* percept (as will this study, largely aligning with Moylan's principles and analytic framework applied to the aural analysis of the recording—hereby referred to as the *sonic*—domain).[8] Finally, Amanda Sewell's (2013, pp. 26–67) useful typology of hip-hop samples across a range of seminal phonographic case studies identifies their structural organisation and aesthetic implications, but not the inner workings of a sample-based record's *mixing* architecture and timbral make-up (essential knowledge if these mechanics are to be deployed in the service of source sample *creation* and, then, *use*).

In other words, whilst Williams (2010, p. 1) and other scholars choose to transcend "narrow discourses of 'sampling'", this examination (re)considers the importance of close readings of content (and process) within the sonic domain, in order to evaluate the character and context of unique (read 'samplable') phonographic ephemera. The aim of the approach, however, is not to champion a return back to an intrinsic flavour of musicology but, rather, to proceed with a phenomenological questioning of aural experiences in sample-based Hip Hop, in order to decipher their underlying causality and poetics (more on this, in the methodological section below). This quest, of course, assumes elevated importance for a practice-based examination, as it transcends description, with the potential to inform—and interact with—*application*. As Albin Zak III (2001, p. 26) has highlighted:

> Recording Practice in-and-of-itself remains stubbornly absent from the lion's share of published research. In my opinion, it will remain absent until a unified "disciplinary" approach to analyzing record making (and not just records) finally emerges, an approach that conceives, and explains, musical practice of recording technology... as musical communication per se.

In order to address the sonic domain as part of this quest, I will engage a range of theoretical perspectives drawn from musicological literature focusing on record production. Simon Zagorski-Thomas's (2014, pp. 37–46) typological categories of "sonic cartoons" (relating to the sonic "signatures" of records), "using technology" (relevant in this context to the interaction of recorded sound with sampling), and "aesthetics and consumer influence" (considering the cultural dimensions of the sound of recorded music and its perception) are helpful starting points to frame the investigation.[9] Furthermore, his notion of "staging" is employed as a catalyst for identifying timbral, spatial, or mechanical statements in recorded sound (Zagorski-Thomas, 2014, pp. 70–91).[10] In the context of sample-based Hip Hop, this research will argue, staging phenomena assume exponential dimensions, and the practice of constructing and then juxtaposing phonographic 'stages' becomes a distinguishing textural characteristic of the project's proposed beat-making poetics. Zak's (2001, pp. 49–96) broad categorisation of recorded sound phenomena into "musical performance", "timbre", "echo", "ambience" (reverberation), and "texture" in *The poetics of rock* draws out a foundational spectrum of practical variables responsible for the construction of unique phonographic qualities. His rich analysis of the effects of recording spaces and vintage equipment is particularly valuable to this investigation, given that the musical content he analyses often forms much of the sampling material in hip-hop production (Zak III, 2001, pp. 97–127).

Additionally, a number of scholars provide illuminating information on the creative use (and abuse) of production technology through discourse on tradition versus innovation (Bennett, 2012; McIntyre, 2015); creative and technical exploration of particular studios and eras (Seay, 2012; Jarrett, 2014); the narrative implications of mixing/staging practice (Liu-Rosenbaum, 2012); and staging as it relates to functional considerations and reproduction media (Zagorski-Thomas, 2009, 2010). This literature, when combined with historical texts on the evolution of recording technologies (Milner, 2009; Horning, 2013), and textbooks on classic or unique recording and production techniques (Cunningham, 1998; Owsinksi, 1999, 2000, 2013; Massey, 2000; Granata, 2003; Stavrou, 2003; Huber and Runstein, 2013; Katz, 2013; Mixerman, 2014a,b; Corbett, 2015; Senior, 2015) offers a rich pool of data on the technical processes deployed, and the factors contributing to *signature sounds*.

Methodology/ies and research design

What I attempt throughout—and outside of—this book is an unapologetic fusion of theorising and beat-making practice. There are two fundamental forces driving this vision. On the one hand, this duality represents the two consistent streams of my professional life (beat-maker and academic). Therefore, I envision the text to remain largely performative, relaying the sonic experiments and creative eureka moments to the best degree possible that (textual) reification may allow (and schematic strategies are also deployed to support the intended 'translation'). The relationship between theory and practice, however, is dynamic, and theoretical analysis is not simply deployed as a means to reflect on and understand practice; instead, it drives, inspires, and modifies practice through a reflexive process of constant conceptual 'contamination'. The associated album of instrumentals simultaneously released with this book is therefore not old work simply being deconstructed, but unfolding praxis having been produced very much progressively and in parallel to the text. The released version of the album, entitled *KATALH3H (Beats Deluxe)*, can be accessed from www.stereo-mike.com, with corresponding chapter playlists available via the 'Album' menu tab.

On the other hand, this intertwined fusion of theoretical analysis and praxis also stems from a career-long obsession to improve the ways in which academia and education deal with the synthesis of the two streams. I have found that students of creative practice degrees overwhelmingly struggle with meaningful analysis, integration of theories, and extraction of systems or concepts out of their artistic work. And they are not to blame. Institutions and curricula suffer from an antithesis of established, singular, and residual methodologies, not only conflicting with the inherent spirit of 'play' in the arts, but also from a somewhat laissez-faire accounting of research methods, without focusing on the ways in which they can be meaningfully synthesised with (read, specifically designed for) different types of creative projects. I do not propose that what this book attempts represents a holistic solution to this duality, but I do hope that this particular design exemplifies some of the ways in which a palette of methodological strategies can be considered, mapped to, and integrated within a creative practice trajectory. Of course, each creative project is unique, and the specific questions, objectives, and criteria will differ. But I expect that as readers follow me on this quest of the phonographic dimensions in sample-*creating*-based Hip Hop, they may also get inspired by some of the methods deployed in integrating theory and praxis. Readers less interested in the underlying design, however, may choose to skip the following paragraphs and fast-forward to the 'Phonographic (aural) analysis and its phenomenological dimensions' section a few pages below. The book's main chapters will make analytical sense in their own right, nevertheless mirroring the methodological exposition expressed next.

There are three methodological frameworks acting as conceptual allies in this undertaking. The first one stems from the question of how do we meaningfully 'look' at personal artistic praxis, and how do we extract useful, and potentially generalisable and transferable 'data' from its unfolding. With a focus on a "researcher's personal experience … [illuminating] the culture under study" (Ellis and Bochner, 2000, p. 740), one of the more potent and applicable forms of auto-methodology or self-study appears to be *autoethnography*, presenting considerable appeal to arts-based researchers investigating wider aesthetic phenomena through the lens of their own practice. But this potential comes with a number of tensions. Having evolved out of ethnography, and representing a post-structural paradigm shift in social anthropology, autoethnography pursues "ethical agenda", expresses "fieldwork evocatively" (Ellis and Bochner, 2006, p. 445), and uses "our experience to engage ourselves, others, culture(s), politics, and social research" (Adams, Holman Jones and Ellis, 2015, p. 2). These are methodological purposes and conditions highlighting that the leap from culture to aesthetics requires some precaution before autoethnography can be congruously applied to contemporary artistic fields. The concern is expressed, for example, in McRae's (2009, p. 143) music autoethnography, where he admits: "I may not implicate myself or my audience in my performance in resistant ways, but my performance does create and perpetuate a certain aesthetic".

Nevertheless, there have been notable attempts to integrate autoethnography into musicological pursuits, resolving some of these tensions, as is evident by a number of collected chapters comprising *Music autoethnographies* (Bartleet and Ellis, 2009). Here, through a variety of foci, the authors link notions such as "image, gesture and style to cultural expression and identity" (Bloustien cited in Scott-Hoy, 2009, p. 50); improvisation to "the harmonisation of one's musical personality with social environment" (George E. Lewis cited in Knight, 2009, p. 80); and instrumental performance to "the historical, social, and cultural meanings that are layered on and in" (McRae, 2009, p. 149) a particular instrument.

Furthermore, and with regard to sonic phenomena, Knight (2009, p. 82) resorts to performative, onomatopoeic inventions to describe the electroacoustic manipulation of his trumpet recordings, offering accounts such as: "thump—wheeze—then process and texturise the sound and listen as it layers up over previously sampled clicks and exhalations".

From a more technical perspective, Anthony (2018) engages a largely autoethnographic approach to analyse popular music sound mixing as performance using hardware studio equipment. And, specifically in relation to Hip Hop, Harrison (2014) deploys an arts-based autoethnographic lens to study underground hip-hop music making—a study that incidentally recounts instances of sampling that take advantage of improvised live performance as source material. Findlay-Walsh (2018, pp. 122–123) expands the reach of the method to *sonic* autoethnographies, "swapping the writing and interweaving of texts for the recording and layering of first-person auditory perspectives"; and seeking "to involve and to implicate subsequent listeners in the enquiry, generating productive tensions between different listening perspectives, as well as between different recorded auditory environments". Findlay-Walsh's approach highlights some of the textual limitations in relaying, integrating, and meaningfully examining other media (here, sonic art), especially when these constitute the primary raw materials of the artform in question. As conductor/autoethnographer Bartleet (2009, p. 715) explains:

> Due to the musical nature of my project, my reflections were not always in text-based formats. I filmed my rehearsals and concerts and undertook interviews with colleagues and used sound recordings and photographic images to reflect on significant moments throughout my musical development.

The practical implication being that effective mechanisms, formats, and technologies need to be carefully considered for the effective observation and capture of artistic praxis, should nuanced processual detail be systematically gathered to fuel ensuing interpretation. Bartleet's approach, furthermore, raises the point that not only a range of formats and technologies, but also different types of methodologies, may require merging in pursuit of reflection (here, fusing ethnographic interviews with self-study).

A related tension lies in the—textual—tone that self-study and related auto-methodologies should assume in pursuit of validity, which is particularly pertinent to arts-based contexts. Proponents of evocative, narrative approaches (especially in autoethnographic enquiry) favour storied writing that embraces subjectivity, vulnerability, and an artistic, performative style of presentation.[11] These are qualities that promote honesty, authenticity, and nuanced detail, offering increased disclosure of the authors' biases and backgrounds within the text. Borne out of the crisis of representation, such autoethnographic accounts engage the reader in a bilateral dynamic, amplifying the emotive effect of lived experiences through performative storytelling, and drawing the reader into a co-experiencing of the expressed epiphanies through empathy.[12]

The counterargument coming from 'analytic' voices (for example: Anderson, 2006) is whether this mode of effective *showing* needs to be met by coherent *telling* (Pitard, 2019, p. 1834), where the exposition of experiences and events is followed through with structured analysis serving particular research contexts. Scholars, therefore, synthesising auto-methodological approaches in various fields, deploy elements of autoethnography (to evoke experiential resonances in a striking fashion), but proceed onto analytical frameworks that help contextualise and generalise these as part of a *theorising*. For example, Gorichanaz (2017, pp. 3–4) admits that a "boundary" between a proposed phenomenological 'auto-hermeneutics' and autoethnography "is admittedly a diffuse one", but indeed fuses elements of the two approaches under a Heideggerian/existential paradigm of "analysis *in situ*, emphasizing the role of interpretation" (my emphasis).[13] The delineation, respectively, is qualified by the focus on phenomena rather than culture, but with the common denominator of a person's "lived experience" often being "inseparable" from culture (Gorichanaz, 2017, p. 4). Pitard (2019, p. 1840), on the other hand, taking a Husserlian/transcendental approach, bridges the phenomenological-autoethnographic gap by deploying 'vignettes' and 'anecdotes' that capture "prereflective stage[s]" (or 'transcendental reductions' in phenomenological parlance), as part of a multi-step analytical framework. The framework progresses through consecutive "layers of awareness that might otherwise remain experienced but concealed, [taking] the reader on a collaborative journey of cultural discovery" (Pitard, 2019, p. 1829). Pitard, thus, manages to successfully merge evocative writing with phenomenological 'variational' questioning, having engineered a design that demands layered interpretation from multiple perspectives in the service of increased reflexivity.[14] As such the "*nuanced*, *complex* and *insider* insights" (Adams, Holman Jones and Ellis, 2015, p. 105, original emphasis) that warrant reflection and ignite reflexivity in autoethnography, parallel the phenomenological "insistence on the importance of carefully attending to the phenomena in their full *concreteness*, [and] the importance of unprejudiced descriptions" (Zahavi, 2021, p. 272, my emphasis). Similarly, rapper-researcher Dave Hook (2020, pp. 73–74) opts for an analytic flavour of autoethnography by building a multi-method "scaffold" around reflexive narratives to "validate and support … findings"—he states that:

> in the case of self-analysis of an artist's work, the autoethnographic process is travelling in the opposite direction to the more sociologically inclined creative-writing format of autoethnography championed by Bochner and Ellis (2003). Rather than being designed to find creative ways of sharing and examining work in a traditionally academic field, it is about allowing the artist to take part in analysis and research of their existing creative output.

The deployment of autoethnography in conjunction with (and in the service of) arts-based enquiry, inevitably raises the issue—and opportunities therein—of merging multiple methods as part of a study, and negotiating their hierarchical interplay. Here, I turn to the second theoretical framework that supports this line of design thinking. The direction of travel on contemporary or practice-based creative research (Collins, 2010; Rogers, 2012; Barrett and Bolt, 2014; Leavy, 2015; Alvesson and Sköldberg, 2018) has been moving away from monological positions, acknowledging the benefits of promoting multi-method integration within the same investigation. Alvesson and Sköldberg (2018), for example, argue that methodological paradigms such as grounded theory, hermeneutics, post-structuralism, and (various types of) ethnography may be apt for *particular phases* of a research examination, rather than pose strict investigative frames within which a whole study should remain contained.

But crucially, the study of one's own creative practice brings to the forefront issues of rigour, value, validity, and quality often fuelled by residual (positivist) critical stances towards any form of auto-methodology (Gorichanaz, 2017). The rationale behind discrediting auto-methodologies can be attributed to the inherently small sample size (the self), and the reduced distance from the subject/area/praxis studied, which—it is inferred—would otherwise help 'objectify' findings (Gorichanaz, 2017). As a result, the onus falls upon the researcher–practitioner to devise effective strategies for the observation, collection, and analysis of self-'data' that are transparent, meaningful, and actionable for ensuing interpretation. Alvesson and Sköldberg (2018) indeed propose that a flexible interpretive approach, meaningfully mapping methods to issues stipulated by the research question(s), is a way to ensure true *reflexivity*. This is a notion that resonates with Cheryl S. Le Roux's (2017, p. 198) examination of rigour, specifically in autoethnographic research, where "the researcher and the researched are often the same person"—therefore "[a]utoethnography as a methodology demands multi-layered levels of researcher reflexivity".

A common denominator that persists throughout the majority of self-study approaches exposed here (phenomenological auto-hermeneutics; evocative, analytic, and/or music/sonic autoethnography) is the pursuit of reflexivity, ensured via the integration of a suitably wide interpretive repertoire (Alvesson and Sköldberg, 2018, p. 250). Given the subjective nature of artistic inquiry, it becomes clear that actioning reflexivity can function both as a validating catalyst *and* a driving force behind multi-layered interpretations of phenomena or experiences. But two final factors—for the purposes of this methodological exposition—require addressing: how to transcend mere reflection over process and achieve critical/true reflexivity, and how to do so via the integration of (artistic) praxis within the research design.

In her award-winning article "On becoming a critically reflexive practitioner", Ann L. Cunliffe (2004/2016) presents a pedagogical framework—the third conceptual ally here—designed for management education, which is potentially transferable to most creative praxis. Cunliffe (2016, pp. 753–754) proposes that "we [can] make sense of experience" through:

a) "reflex interaction" (action based on "instinct, habit, and/or memory");
b) "reflective analysis" (thinking about, categorising, and explaining an object, "often using theory to help us see our practice in different ways"); and
c) "critically reflexive questioning" (drawing "on social constructionist assumptions to highlight subjective, multiple, constructed realities" and "exposing contradictions, doubts, dilemmas, and possibilities").

Cunliffe's triangle of experiential sense-making mirrors Pitard's multi-stage analytical model, offering a malleable structural purview over reflexive multi-methodological designs such as those promoted by Alvesson and Sköldberg (2018). Figure I.1 provides a schematic representation of the iterative phases in this research design, mapped against relevant methodological paradigms and Cunliffe's triangle of sense-making stages. The following section discusses this bricolage model as it is actioned—in different proportions—throughout this book.

Reflex interaction, (grounded theory) and creative praxis

In this project, the praxis of both creating source audio that facilitates a sample-based creative process, and the sample-based process itself, offer rich opportunities for *reflex interaction*: using tacit knowledge as a musician, studio engineer, and record producer to create original audio content over these two different stages. Habit, memory, and previous experience as a professional practitioner inform the progression from initial musical ideas

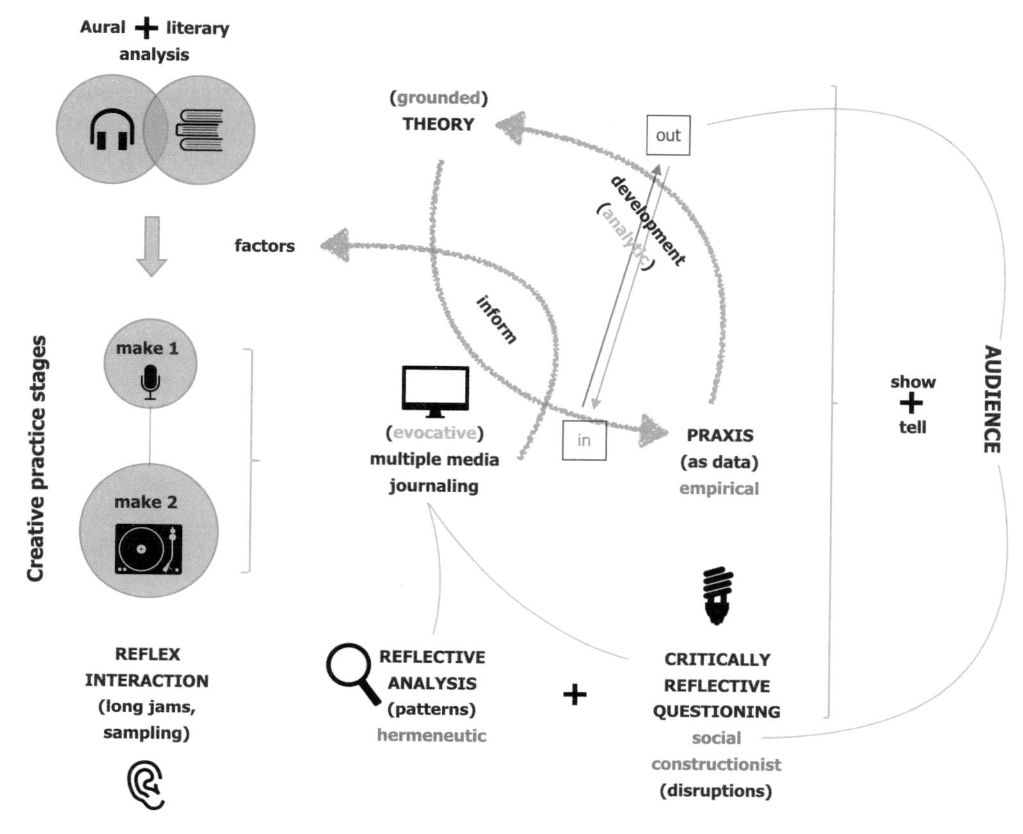

FIGURE I.1 A schematic representation of actioned phases in the research design, mapped against fitting methodological paradigms and (Cunliffe's triangle of) sense-making stages.

(inception) to developing sonic outputs (production). At source level, this engagement takes the form of responding to abstract musical ideas; sonifying them through instrumental performance; capturing these sonifications with recording technologies; layering further musical ideas (which follow the same progression from abstraction to sonic materiality); organising structure and form; and shaping and processing the resulting sound through studio haptics and technologies (or their emulation thereof via computer operating scripts). At the sample-based level, the engagement takes the form of selecting audio segments (*digging*), further sonic manipulation, and a percussive style of improvisation, performance, and programming/composition resulting in reimagined musical phrases (*chopping*/beat-making).

As such, the creative practice experiment facilitates the generation of data, echoing an empirical, grounded theory (inductive) approach. However, as the investigation poses pertinent questions about:

- what is 'phonographic' at the source level of sample-based Hip Hop,
- the aesthetic implications of the past in its sonic manifestation, and
- the material poetics of inter-stylistic synthesis and phonographic juxtaposition,

it also makes sense to inform practice with theoretical preunderstandings. These can take the form of findings drawn from literature, textbooks, and discography that respectively guide the production process through a constructivist accumulation of historical, musicological, and technical detail on sampling processes; past phonographic workflows and technologies; as well as phonographic referencing (through aural analysis for the latter). Figures I.2 and I.3 provide a schematic representation of the phases described.

Phase 1

The first phase of the research design involves the analysis of discography, historiographical and musicological literature on the notion of phonographic signatures, as well as literature on sampling practices. The aim is to identify the sonic factors that draw the sample-based producer into the selection of particular audio sources/segments. These factors are pursued and recreated in the following, applied creative-practice phase, and further informed by reflective analysis facilitated by multi-media journaling (as will be discussed in the following section).

Phase 2

The primary stage of the second—applied—phase incorporates the recording, mixing, and mastering of original content, referencing classic phonographic styles and eras. This musical content is performed, composed/improvised, and engineered with the aim of providing a rich pool of raw sonic material for the subsequent, sample-based stage. The production of the content is informed by historical and technical detail derived from the studied range of historiographical and musicological sources, audio engineering and production textbooks, and aural analysis of discography. The second stage consists of the selection of numerous (typically short) samples from the recordings/masters made in the previous stage, assigned

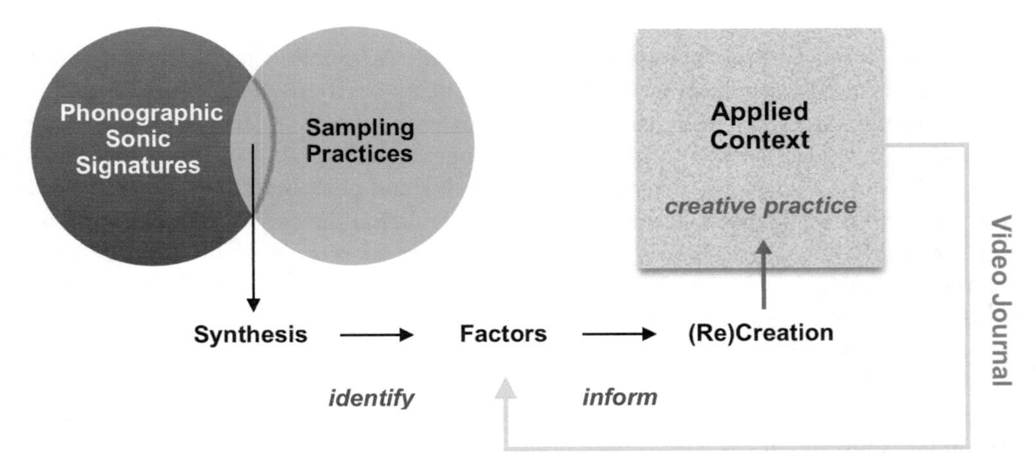

FIGURE I.2 First phase of the research design depicting the deployment of literature on phonographic signatures and sampling practices to identify key sonic factors influencing sample selection. These factors inform the creation of original source audio in the following, applied creative-practice phase.

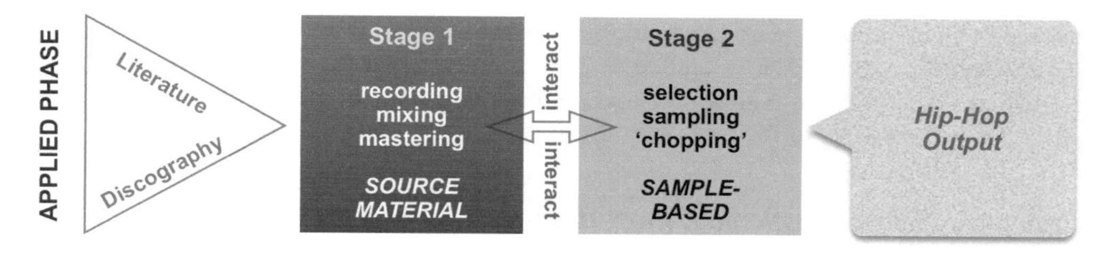

FIGURE I.3 Second—applied—phase of the research design comprising two practice-based stages. The primary stage involves the recording, mixing, and mastering of original content, informed by historical and technical detail derived from the first phase. The second stage involves the selection of samples from the recordings/masters made in stage one, facilitating the beat-making process that leads to the associated album.

as audio segments to the drum pads of typical sampling drum machines used in the genre: the process promotes a percussive style of improvisation, performance, and composition resulting in reimagined musical phrases exemplifying the sample-based hip-hop aesthetic (beat-making). The artefacts of this stage are, in turn, mixed as finalised productions (making up the end practice-based outputs—the album beats—of the wider project).

Reflective analysis, (hermeneutics) and journaling

The next step in the cycle is to reflect on ongoing practice (and interim outputs) via the use of further theory, reflectively analysing the practice/experience to 'see' or interpret it in developmentally informed ways. *Reflective analysis* here takes a hermeneutic look at 'texts' created out of the practice: these include textual journaling, as well as the capture of praxis through various media (photography, sonic archiving, video footage).[15] Here, patterns emerge out of the tacit reflexes, and significant events are recorded (and acknowledged) in the praxis: what autoethnographers refer to as "aesthetic moments" (Adams, Holman Jones and Ellis, 2015, p. 49), "epiphanies", or "transformative experiences" (Ellis, Adams and Bochner, 2011, p. 275). From an autoethnographic perspective, the creative work in its natural context (studio/technologies) functions as fieldwork, and the varying media used to capture and initially reflect on it offer a range of opportunities for interpretation and "thick descriptions" (Ellis, Adams and Bochner, 2011, p. 277). Textual journaling, for instance, inherently offers an initial level of interpretation as an active recording of events, while photography and video footage represent rich, if more passive (pre-reflective) recordings of events, which in turn require interpretation, description, and added narrative. Autoethnography opts for a 'thematising' of narratives, rather than a strict coding of data, which Adams, Holman Jones and Ellis (2015, p. 77) describe as a search "for *clues*: repeated images, phrases, and/or experiences ... [which] helps us imagine a logic or pattern to our narrative and to explicitly connect personal experience with culture" (original emphasis). This form of thinking-in-action has notable benefits when interjected with practice, allowing new theoretical understandings to emerge and positively affect further praxis.

It is also possible to conceive of the first creative practice stage in this design as a process of construction that itself carries pre-understandings due to its intentionality; and the second, sample-based stage as an opportunity for *sonic interpretation* (testing and using the content produced in the first stage within a specific stylistic frame). Furthermore, the possibilities offered by contemporary music technology in terms of sonic archiving mean that, from a musicological perspective, it is feasible to closely study the parts, elements, or layers that contribute to the whole (record or phonographic moment), mirroring a hermeneutic, if rather intrinsic, paradigm.[16]

Critically reflexive questioning, (post-structuralism) and the 'outside'

However, the limitation of the hermeneutic approach is the pursuit of a single, underlying, and unifying interpretation. In the context of a self-study this may minimise the verisimilitude of investigative findings, as the interpretive effort can be limited by a sole practitioner's artistic frame. This is where the pluralism of a *critically reflexive questioning* may be of benefit. This multiplicity is expressed in two fundamental ways in this design. Firstly, in the cyclical interplay of theoretical development and creative praxis, developing theories and assumptions become *tested* in what Ihde (2012, p. 108) describes as artistic—resembling phenomenological—*playfulness*: "In actually exercising fantasy variations, the arts echo the Aristotelean dictum that poetics is ultimately more true than history. It is out of possibility that the undiscovered is found and created". Or in Cunliffe's (2016, p. 751) words: "instead of applying theory to practice, critical reflexivity emphasizes praxis—questioning our own assumptions and taken-for-granted actions". Producing music within a stylistic frame alone constitutes a *dialogue* with members of an invisible cultural network.[17] These include past and present peers perceived as musical influences, and their works and practices functioning as points of reference and inspiration. Small (1998, pp. 203–204) argues that "since style is concerned with the way in which things relate, it is itself a metaphor for the way in which the society conceives of the pattern which connects".

A second, more direct strategy is to seek peer or audience interaction, feedback, and involvement, in response to developing findings and outputs. Examples may include creative collaboration, public performances or dissemination of audio(visual) works[18]; presentations of findings at academic conferences; and the publication of interim texts communicating ongoing research.[19] Further examples here include collaboration with other artists in the recording studio[20]; and semi-structured interviews with peers, which have unveiled a level of personal and technical detail that would be unlikely to acquire from aural analysis alone (at times correcting assumptions drawn and offering nuanced geo-cultural rationale for essential innovations leveraging inter-stylistic syntheses).[21] These have provided holistic understandings of contextual and cultural factors contributing to unique production signatures.

Phonographic (aural) analysis and its phenomenological dimensions

A crucial tool deployed in the analysis of the sonic domain throughout this book that cannot be taken for granted is the process of *listening*. In *Recording analysis: How the record shapes the song*, Moylan (2020, pp. 59–60) acknowledges that "listening is personal", so it is important to consider more explicitly how personal observations and evaluations of aural phenomena can illuminate wider cultural practices (of the imagined beat-making community). In resonance with previously discussed precautions relating to auto-methodologies, a systematic exposition of listening as an analytical process can be beneficial for the transferability of findings. In *Recording analysis*, Moylan (2020) offers an analytical framework that adopts a phenomenological systematicity towards describing, deconstructing, and explaining aural phenomena contained in phonographic records through the listening experience. Moylan here adheres to fundamental rules of the phenomenological method—attending to phenomena of experience as they appear (sound); beginning with description, not explanation; avoiding hierarchical assumptions or judgements; and drawing out "*essential*", "*structural features* or *invariants* within phenomena"—to identify patterns across multiple phonographic examples (Ihde, 2012, pp. 2–22, original emphasis). He does so by breaking down all aspects of phonographic content into music (elements), lyrics, and recording (material attributes) domains, and these, further, into elements and material content that become the subject of focused study. He adopts visual representations graphing these materials and their interaction as a means to capture nuanced aspects of the audio phenomena, "crystallise" them (Moylan, 2020, p. 338), and bring them closer to textual/visual reification for critical examination. Finally, he stresses the importance of specifying the analytical vantage point and choice of domain-focus as part of the musicological interpretation. In this way, Moylan's 'personal listening' intent functionally satisfies Ihde's (2012, pp. 27–31) "reflexive move":

> Analysis moves from that which is experienced toward its reflexive reference in the how of experience, and terminates in the constitution of the "I" ... the phenomenological "I" takes on its significance through its encounter with things, persons and every type of otherness it may meet.

The chapters comprising the main body of this text adopt this analytical paradigm to draw out the material *invariants* contributing to the sample-based aesthetic, with a focus on elements of the sonic domain—taking a variety of 'probing' approaches facilitated by the multi-method opportunities contained within the bricolage design. As such, the stance assumed towards knowledge (articulated or emergent) subscribes to Kincheloe's *critical constructivist* position that it "is temporal and culturally situated" (Kincheloe 2005a, cited in Rogers, 2012, p. 10). It also aligns with Kincheloe's "symbiotic hermeneutics" approach, which according to Rogers (2012, p. 10) "means that bricoleurs seek out ways that phenomena are interconnected with other phenomena, and socially constructed in a dialogue between culture, institutions, and historical contexts". In all, the bricolage design here synthesises autoethnography, hermeneutics, ethnographic strategies (interviews), aural and literary analysis, and arts-based inquiry to facilitate a phenomenological questioning of phonographic phenomena in the context of sample-*creating*-based Hip Hop.

Structural organisation

The book's chapters deploy bespoke combinations of the methodological strategies comprising this bricolage palette, in order to deal with the respective foci from a number of perspectives and offer triangulation (or *variations,* in the phenomenological sense). A crucial catalyst for the theoretical positions that will emerge, remains my specialisation as an audio engineer, providing a (trained) *mixing* lens/ear to the sample-based investigations and phonographic deconstructions that will follow.[22] Mixing craft, in the case of sample-based musics, constitutes much more than a post-production process: it is an essential praxis responsible for, and situated at the heart of, their inception and creation. It is also the means by which the sonic character of their outputs and unique timbral footprint are negotiated, sculpted, and manifested. The materials 'jammed' within sample-based practice may appear like motifs, 'hooks', or phrases to the listener, but they really are the perceived music-language abstractions emanating out of the interplay of phonographic objects. These objects, the book will demonstrate, are in themselves complex material entities rich in elements pertaining to the sonic domain—their poiesis (manipulation, interaction, and juxtaposition), resulting in aural phenomena of sublime aesthesis, both for makers (inspiring creation) and listeners (enticing reception). In pragmatic contexts, where the raw materials necessitate a *reimagining*, the study of the underlying mechanics can empower informed creativity and, potentially, *innovation*. As such, the conceptual framework that emerges for a reimagining of sample-based Hip Hop—and which underpins the evolving practice and remainder chapters—can be summarised under the following pre-understandings:

i. Sample-*creating*-based practice is part of a contemporary artistic-cultural zeitgeist—responding to crisis (necessity) with (creative) synthesis
ii. Beat-making is a manifestation of 'play' with sonic objects
iii. Source objects in sample-based Hip Hop are constructs communicating phonographic context, typically characterised by past sonic signatures
iv. The appeal of the sample-based aesthetic is the result of a juxtaposed phonographic poiesis, interacting with previously crystallised phonographic poetics
v. Newly recorded live performance benefits from a sonic 'distancing', which can be expressed through mixing practices that imbue spatial and temporal qualities in their *staging*
vi. The staging manifestations of sonic-domain relationships are expressed as material ephemera that facilitate a fertile palette of creative opportunity for sample-based music making.

Chapter 0 functions as a juncture between this introduction and the main-body chapters of the text, firmly placing myself as the author, at the centre of the phenomenon. Having come face-to-face with the described conundrum in the context of a career as professional musician, the realisation is narrated as a tale of pragmatic ultimata requiring resolution. The narrative leads to an interpretation of the experience as an opportunity for a practice-based examination of alternative sample-based approaches (the *need-for-study* area), recognising their implications for a wider creative community. Sample-*creating*-based practice is therefore deliberately framed as a response to this arising necessity, and its manifestation explored as part of wider contemporary artistic trends. The chapter addresses the geographical remit of the project, contextualising my experience as a European hip-hop artist (a UK-based beat-maker rapping in Greek), who sonically communicates through—and creates from within—an African-American stylistic lexicon. This frame presents an opportunity to examine the implications of the contemporary sample-licencing landscape on beat-making creative practices from a universal, diasporic lens. Opting for autoethnography (and a neo-narrative writing ethos), the interpretation of personal studio practice rendered as fieldwork enables the extraction of developmental interpretations from the textual (journal), sonic (archiving), and (making-of) video data collected. The process, in turn, leads to reflexive understandings, connecting the insider practice to the larger aesthetic phenomenon. As Guthrie P. Ramsey, Jr. states in *Race music* (2003, p. 22), one of the goals of such projects is "to have readers understand something about some of the sources and grounding of [the author's] own critical voice and biases".

Chapters 1–6 thematically encompass the sequence of creative stages typically comprising a music production trajectory—and the specific technologies deployed—with respective foci on:

1. (inter-stylistic) *composition*;
2. (characteristic) *tools* facilitating hip-hop composition;
3. (reverse) *engineering*;
4. (the 'magic' of sample-based) *production*;
5. *mixing* (records within records); and
6. a form of (exponential) re/*mastering*.

In Chapter 1—through the merging of literary analysis, phonographic case studies (enriched by semi-structured interviews with practitioners), and own creative practice—the bricolage approach takes *Blues Hop* as a subgenre case-in-point: the objective is to illustrate how sonic pursuits, informed by a sample-based aesthetic context, drive compositional innovation (with motivic, harmonic, and *textural* implications). Structurally, the phonographic case studies assume an inverse probing direction, starting with the analysis of music domain utterances, to expose their underlying sonic rationale and hidden dimensions. This exposition aims to reveal the compositional, performative, and (inter-)stylistic resonances of beat-making, when understood as a creative form that prioritises sonic objects in its interplay. The interviews carried out with remixer extraordinaire Amerigo Gazaway and rapper Abdominal (and The Obliques) illuminate the phonographic analysis with insider knowledge (a Salaam Remi production for a Nas track is also put under the aural microscope); while the ensuing deconstruction of an original blues composition written to provide the raw sonic content for subsequent beat-making is added to the mix. Accompanied by work-in-progress soundbites extracted from sonic archiving, the musicological findings are expressed in both sonic and textual terms, amplifying the performability of the process(es), and demonstrating the Afrological dimensions of the percussive reimagining of multitrack content via the use of sampling drum machines. Furthermore, the chapter extrapolates on the cross-genre implications of the, a priori, inter-stylistic workflows discussed.

At this point, it becomes important to examine the interrelationship between the tools frequently (and historically) deployed in sample-based hip-hop record production and the ensuing aesthetic implications their operating scripts and physical interfaces have had upon workflows and resulting production *signatures*. Using the Akai MPC range of sampling drum machines as the main focus in Chapter 2 (rationalised as such due to the contemporary development of the technology alongside the boom-bap—or sample-based—aesthetic), a typology of sonic signatures is extracted and mapped to the technology's affordances, further illustrated by key phonographic releases designating hip-hop eras and subgenres.[23] The boom-bap sound is traced from its origins in the mid- to late-1980s, through to its current use as an East Coast production reference, and the findings from a number of representative case studies form a systematic typology of technical characteristics correlated to creative approaches and resulting production traits. The discussion informs speculation about the future of the MPC, its technological descendants, and the footprint of its aesthetic on emerging styles and technologies. The chapter also facilitates a better understanding of the technological nuance necessary to observe the analysis of creative processes in the following chapters.

As sample-based practitioners have been pursuing alternative routes towards music creation, including the recording of live instrumentation and the production of intermediate sampling material, it is important to consider the variables that enable an effective interaction between original source content and the hip-hop process. Chapter 3 addresses source objects deployed in sample-based Hip Hop as constructs that communicate phonographic context, typically characterised by past sonic signatures. It proposes that Hip Hop's 'meta' aesthetic is borne out of the fusion of sampling processes and phonographic signatures, examining the bi-directional dynamic involved in their (re)construction, and questioning the genre's complex relationship with the past. Covering, but also expanding beyond, a deterministic approach to re-engineering that classifies signal flow variables, the chapter problematises the notion of phonographic context, extending the understanding of record production as—a form of material—composition. Four aesthetic deductions form the main arguments of the text (the function of nostalgia, the amount of historicity required, the notion of phonographic 'magic', and the irony of reconstruction), paving the way for the foci of the following chapters; and drawing parallels between the reconstructive phenomena in hip-hop practice discussed and a wider *metamodern* "structure of feeling" (Vermeulen and Van Den Akker, 2010) observed in contemporary culture.

Since rap producers attribute an inherent 'magic' to working with past phonographic samples and fans appear spellbound by the resulting sonic collage, Chapter 4 examines the music's unique recipe of phonographic juxtaposition. It does so by exploring the conditions of this ascribed 'magic', investigating gaps in perception between emotional and intellectual effect, and deciphering parallels in the practice and vocabulary mobilised against a range of genres in performance magic. The chapter traces the appeal of the sample-based aesthetic in the creative and performative interplay between multiple levels of phonographic poetics crystallised in material (sonic domain) form. By taking a systematic approach to deconstructing examples from discography and blending the aural analysis findings with practice-based investigations, it illustrates—via schematic representations—exponential staging phenomena recognised as essential for the music's mesmerising effects. The notion of staging is therefore extended to cover the striking juxtaposition of spatial illusions taking place in sample-based record production.

By looking at sample-based record production through the lens of "meta-music (music about music)" (Mudede, 2003), Chapter 5 amplifies the multitude of material implications this understanding has for the musicological study of sample-based Hip Hop. Therefore, the chapter questions what renders a sampled source into a phonographic *object*—a phonographic 'other'—that is aesthetically desirable for, and usable in, the context of hip-hop record production: what are the mechanisms, processes, and practices that infuse sonic signatures of phonographic *otherness* onto newly created objects, and how can this 'otherness' be defined? The chapter acknowledges that newly captured live performance benefits from a sonic 'distancing' or infused *alterity*, and explores how this quality may be expressed through mixing practices that consciously imbue spatial and temporal dimensions in their staging. Synthesising the technical with the aesthetic, the chapter deciphers the exponential staging phenomena situated at the heart of how this 'otherness' is negotiated (and constructed) in practice. Sections focus on the spatial-textural continuum, the sonic draw of samples, as well as notions of multi-dimensionality, juxtaposition, and additive processes in sample-based music making. The gap between live performance and the phonographic sample is re-addressed, a sample's 'aura' deconstructed, and the notion of making records—*not recordings*—within records further exposed.

The final chapter attributes the appeal of past phonographic signatures also to *mastering* practices, extending the investigation beyond the recording and mixing realms, and deconstructing how their sonic manifestations interact with beat-making. The chapter proposes that the staging manifestations of sonic-domain relationships materially crystallised within phonographic *masters* present a fertile spectrum of malleable variables in the hands of beat-makers. In this context, the lesser attention given to the sonic 'object' calls for a focused examination of the specific variables involved in the fusion of 'past' (or previously constructed), and present phonographic processes. This inquiry focuses on the merging of 'staging' illusions as a subset of such variables, questioning how full-range masters function as source content in sample-based engineering and production practices. The examination explores how hip-hop producers negotiate the dimensions of 'depth,' 'height,' and 'width' imbued into audio masters when used as sampled sources; but also the ways in which beat-makers stage previously-constructed mix architectures into newly-juxtaposed sonic illusions.

At the end of each chapter, a recommended chapter playlist (in order of track appearance in the text) outlines the beats discussed, deconstructed, exemplified in, or informed by the text, providing sonic summaries of the theoretical notions manifested in the associated album. Some of these tracks appear in more than one chapter, as they express multiplicities of theoretical notions or expositions, and contain sonic elements that warrant discussion in multiple sections. Put on some headphones (or turn on your favourite speakers), and enjoy the ride!

Notes

1 Steve Collins (2008) reports that the "aftermath" of cases such as the 1991 lawsuit involving Biz Markie's 'Alone Again' from *I Need a Haircut* (1991) "sounded a death knell for unlicensed sampling", causing "substantial changes in appropriative music", and resulting in "a stringent … licensing system to govern the practice". He adds that: "The sampling musician is subject to the arbitrary licensing fees demanded by the copyright owner" because "there is no compulsory licence for sound recordings and therefore no statutory controlled royalty rates"; and offers Kanye West's 'All Falls Down' from *The College Dropout* (2004), as an example of a hip-hop production opting for a re-recording (featuring Syleena Johnson), instead of paying $150,000 for a license of the intended original (by Lauryn Hill) (Collins, 2008).

2 By 'mechanical borrowing' I am referring to the use of *recorded sound* from a previously released music recording, which is subject to mechanical copyright (PRS for Music, 2021).

3 'Chopping' and 'chopped' are commonly used terms in hip-hop production terminology indicating processes creatively utilising and re-arranging edited or truncated audio segments, often digitally sampled from phonographic masters. Kajikawa (2015, p. 164) defines 'chopping' as "the process of dividing a digital sample into any number of smaller parts and rearranging them to create a new pattern".

4 Schloss (2014, p. 79) defines "digging in the crates" as the process of "searching for rare records", which provide "the raw material for sample-based hip-hop" alongside associated functions, such as: "manifesting ties to hip-hop deejaying tradition, 'paying dues', educating producers about various forms of music, and serving as a form of socialization between producers".

5 The reference to *analogue* eras pertains to the use of predominantly analogue technologies and media, which are understood to infuse particular and recognisable sonic characteristics to recorded music, signifying specific periods in record production history (see, for example: Bennett, 2012).

6 For example, identifying within a sample a distinct performance by a particular drummer (such as Clive Stubblefield, Gregory Coleman, or John Bonham); as opposed to perceiving a sample as taken from 1960s funk/soul or 1970s rock discography.

7 Kulkarni (2015, p. 78) defines the Golden Age as "an era in which sampling hit a dizzying new depth of layered complexity and innovation" and "a sublime 10-year period from 1988 to 1998 in which hip hop was artistically more free than it had ever been before".
8 This delineation is a subtle but important one for this text, highlighting that the sonic characteristics of phonographic content are the result not only of recording processes, but of the music production chain as a whole, which also includes *mixing*, *mastering*, and physical *manufacturing* practices.
9 Zagorski-Thomas (2014, p. 68) defines 'sonic signatures' as characteristic sounds that "can relate to particular types of performance or programming characteristics … to spatial characteristics, to particular types of distortion, to the characteristics of particular types of sound sources or instruments or to the type of processing".
10 A number of authors (for example: Lacasse, 2000; Zagorski-Thomas, 2009, 2010; Liu-Rosenbaum, 2012; Holland, 2013; Moylan, 2014/1992) have theorised on the placement of musical elements within the space of a popular music mix, and the concept of 'staging' has emerged as a useful theoretical notion: in essence, it suggests conceptualising a music mix as a 'stage' where the placement—but also the dynamic movement and manipulation—of musical elements (mediation) has thematic and narrative implications (meaning) for both listeners and producers.
11 Adams, Holman Jones, and Ellis (2015, pp. 1–2) define autoethnographic narratives as: "stories of/about the self told through the lens of culture. Autoethnographic stories are artistic and analytic demonstrations of how we come to know, name, and interpret personal and cultural experience".
12 The crisis of representation refers to a "crisis which arises from the (noncontroversial) claim that no interpretive account can ever directly nor completely capture lived *experience*" (Schwandt, 2014, p. 45, original emphasis).
13 The Heideggerian/existential paradigm stands in contrast to a Husserlian/reflective "analytical isolation of phenomena" (Gorichanaz, 2017).
14 'Variational' questioning refers to an active "probing activity" phenomenologists describe as "*variational method*" (Ihde, 2012, p. 23, original emphasis).
15 Hermeneutics is an interpretive paradigm preoccupied with an intuitive revealing of underlying, coherent meaning—or disclosure of truth—hidden within (initially, religious) texts; relying on intuition rather than rigid interpretive rules (and pursuing understanding rather than explanation), hermeneutics questions the text to relate partial or *pre*-understandings to the *whole*, deploying empathy and considering context in order to understand the producers of the texts potentially better than they understand themselves (Alvesson and Sköldberg, 2018, pp. 91–106).
16 An approach Harrison (2014, pp. 10–12), for example, resorts to in his arts-based autoethnography of hip-hop song making. This project deploys a bespoke track/file-naming scheme as part of its archiving strategy, capturing detailed information about the signal flows pertaining to, and the range of processing—serially—applied to, audio recordings (see Chapter 6).
17 Williams (2010, p. 27) refers to this as an "imagined community", while Ramsey, Jr. (2003, p. 4) describes such cultural spaces as "community theaters".
18 Edited segments of the project's video journal, for example, are public-facing as part of the #HipHopTimeMachine vlog—see Video link I.1 available from the 'Book > Videos' menu tab at www.stereo-mike.com.
19 Chapters 1–6 comprising the main body of this text (as well as auxiliary texts referenced throughout) have been published in earlier form as articles or chapters in journals or edited book collections following conference presentations; these constitute outcomes of a strategy to produce research outputs *during* the project and benefit from peer feedback and ongoing discourse *as part* of the bricolage design.
20 Specifically, in the guise of: songwriting and producing original content with musical partners Jo Lord and three-piece band FET 47; and recording and mixing for Latin band Sarabanda and Congolese band Grupo Lokito (see *Bomoko*, 2022). In all of these scenarios, I have previously agreed access to source material for sampling purposes (in the case of the latter bands, in exchange for providing engineering services).
21 See, for instance, Amerigo Gazaway's retelling of the software-assisted extraction of BB King bassist's rhythmic signature to trigger synthetic sub-bass in Chapter 1.
22 Ihde (2012, p. 95) argues for a pragmatic post-phenomenology that interacts with other disciplines so as to offer verticality of variational method:

> to be informed, phenomenology must necessarily rely upon other disciplines. Its view of these disciplines, and particularly its interpretation of what they are doing, may be widely different from what those within the disciplines interpret their task and method to be, but without

these other disciplines, phenomenology would be restricted to the realm of first-person experience. Intersubjective phenomenology is necessarily interdisciplinary phenomenology.

23 Boom Bap is a subgenre/style of Hip Hop, referring onomatopoetically to the sound and rhythm of a heavy bass drum and snare (generally over sparse, sample-based instrumentation characterised by lo-fi sonic qualities). Typically, a sampled break-beat would be supported by synthetic kick- and snare-drum layers, frequently courtesy of a Roland TR-808 drum machine. Chapter 2 delves deeper into the style and its relationship with sampling technologies.

Bibliodiscography

9th Wonder (2021) *SAMPLING IS DEAD.....*, *Instagram*. Available at: https://www.instagram.com/p/CTrus8wgcQk/ (Accessed: 15 September 2021).

Adams, T.E., Holman Jones, S. and Ellis, C. (2015) *Autoethnography: Understanding qualitative research*. New York: Oxford University Press.

Alvesson, M. and Sköldberg, K. (2018) *Reflexive methodology: New vistas for qualitative research*. 3rd edn. Los Angeles, CA: Sage.

Anderson, L. (2006) 'Analytic autoethnography', *Journal of Contemporary Ethnography*, 35(4), pp. 373–396.

Anthony, B. (2018) 'Mixing as a performance: Educating tertiary students in the art of playing audio equipment whilst mixing popular music', *Journal of Music, Technology & Education*, 11(1), pp. 103–122.

Barrett, E. and Bolt, B. (eds) (2014) *Practice as research: Approaches to creative arts enquiry*. London: I.B.Tauris.

Bartleet, B.-L. (2009) 'Behind the baton: Exploring autoethnographic writing in a musical context', *Journal of Contemporary Ethnography*, 38(6), pp. 713–733.

Bartleet, B.-L. and Ellis, C.S. (eds) (2009) *Music autoethnographies: Making autoethnography sing/making music personal*. Bowen Hills: Australian Academic Press.

Beastie Boys (1989) *Paul's Boutique* [CD, Album]. US: Capitol Records, Beastie Boys Records.

Bennett, S. (2012) 'Endless analogue: Situating vintage technologies in the contemporary recording workplace', *Journal on the Art of Record Production*, 7.

Biz Markie (1991) *I Need a Haircut* [Vinyl LP]. Europe: Warner Bros. Records.

Bochner, A.P. and Ellis, C. (2003) 'An introduction to the arts and narrative research: Art as inquiry', *Qualitative Inquiry*, 9(4), pp. 506–514.

Collins, H. (2010) *Creative research: The theory and practice of research for the creative industries*. Lausanne: AVA Publishing.

Collins, S. (2008) 'Waveform pirates: Sampling, piracy and musical creativity', *Journal on the Art of Record Production*, 3.

Corbett, I. (2015) *Mic it! Microphones, microphone techniques, and their impact on the final mix*. Oxon: Focal Press.

Cunliffe, A.L. (2016) 'Republication of "on becoming a critically reflexive practitioner"', *Journal of Management Education*, 40(6), pp. 747–768.

Cunningham, M. (1998) *Good vibrations: A history of record production*. 2nd edn. London: Sanctuary Publishing.

Ellis, C., Adams, T.E. and Bochner, A.P. (2011) 'Autoethnography: An overview', *Historical Social Research/Historische sozialforschung*, 36(4), pp. 273–290.

Ellis, C. and Bochner, A. (2000) 'Autoethnography, personal narrative, reflexivity: Researcher as subject', in N.K. Denzin and Y.S. Lincoln (eds) *Handbook of qualitative research*. 2nd edn. Los Angeles, CA: Sage, pp. 733–768.

Ellis, C.S. and Bochner, A.P. (2006) 'Analyzing analytic autoethnography: An autopsy', *Journal of Contemporary Ethnography*, 35(4), pp. 429–449.

Findlay-Walsh, I. (2018) 'Sonic autoethnographies: Personal listening as compositional context', *Organised Sound*, 23(1), pp. 121–130.

Goldberg, D.A.M. (2004) 'The scratch is hip-hop: Appropriating the phonographic medium', in R. Eglash, J.L. Croissant, G. Di Chiro and R. Fouché (eds) *Appropriating technology: Vernacular science and social power*. Minneapolis: University of Minnesota Press, pp. 107–144.

Gorichanaz, T. (2017) 'Auto-hermeneutics: A phenomenological approach to information experience', *Library and Information Science Research*, 39(1), pp. 1–7.

Granata, C.L. (2003) *Sessions with Sinatra: Frank Sinatra and the art of recording*. Chicago, IL: Chicago Review Press.

Grupo Lokito (2022) *Bomoko* [Vinyl LP]. UK: Malecon Productions.

Harkins, P. (2008) 'Transmission loss and found: The sampler as compositional tool', *Journal on the Art of Record Production*, 4.

Harrison, A.K. (2014) '"What happens in the cabin...": An arts-based autoethnography of underground hip hop song making', *Journal of the Society for American Music*, 8(1), pp. 1–27.

Holland, M. (2013) 'Rock production and staging in non-studio spaces: Presentations of space in Left or Right's 'Buzzy'', *Journal on the Art of Record Production*, 8.

Hook, D. (2020) 'Growing up in hip hop: The expression of self in hypermasculine cultures', *Global Hip Hop Studies*, 1(1), pp. 71–94.

Horning, S.S. (2013) *Chasing sound: Technology, culture, and the art of studio recording from Edison to the LP*. Baltimore, MD: JHU Press.

Huber, D.M. and Runstein, R.E. (2013) *Modern recording techniques*. New York: Routledge.

Ihde, D. (2012) *Experimental phenomenology: Multistabilities*. 2nd edn. Albany: State University of New York.

Jarrett, M. (2014) *Producing country: The inside story of the great recordings*. Middletown, CT: Wesleyan University Press.

Jay-Z (2001) *The Blueprint* [CD, Album]. US: Roc-A-Fella Records.

Kajikawa, L. (2015) *Sounding race in rap songs*. Oakland: University of California Press.

Katz, B. (2013) *Mastering audio: The art and the science*. 2nd edn. Oxon: Focal Press.

Knight, P. (2009) 'Creativity and improvisation: A journey into music', in B.-L. Bartleet and C. Ellis (eds) *Music autoethnographies: Making autoethnography sing/making music personal*. Bowen Hills: Australian Academic Press, pp. 73–84.

Krims, A. (2000) *Rap music and the poetics of identity*. Cambridge: Cambridge University Press.

Kulkarni, N. (2015) *The periodic table of Hip Hop*. London: Random House.

Lacasse, S. (2000) *'Listen to my voice': The evocative power of vocal staging in recorded rock music and other forms of vocal expression*. Unpublished PhD thesis. University of Liverpool.

Le Roux, C.S. (2017) 'Exploring rigour in autoethnographic research', *International Journal of Social Research Methodology*, 20(2), pp. 195–207.

Leavy, P. (2015) *Method meets art: Arts-based research practice*. New York: Guilford Publications.

Liu-Rosenbaum, A. (2012) 'The meaning in the mix: Tracing a sonic narrative in 'When The Levee Breaks'', *Journal on the Art of Record Production*, 7.

Marshall, W. (2006) 'Giving up hip-hop's firstborn: A quest for the real after the death of sampling', *Callaloo*, 29(3), pp. 868–892.

Massey, H. (2000) *Behind the glass: Top record producers tell how they craft the hits*. San Francisco, CA: Backbeat Books.

McIntyre, P. (2015) 'Tradition and innovation in creative studio practice: The use of older gear, processes and ideas in conjunction with digital technologies', *Journal on the Art of Record Production*, 9.

McRae, C. (2009) 'Becoming a bass player: Embodiment in music performance', in B.-L. Bartleet and C. Ellis (eds) *Music autoethnographies: Making autoethnography sing/making music personal*. Bowen Hills: Australian Academic Press, pp. 136–150.

Milner, G. (2009) *Perfecting sound forever: An aural history of recorded music*. New York: Faber and Faber, Inc.

Mixerman (2014a) *Zen and the art of mixing*. 2nd edn. Milwaukee, WI: Hal Leonard Books.

Mixerman (2014b) *Zen and the art of recording*. Milwaukee, WI: Hal Leonard Books.

Moore, A. (2002) 'Authenticity as authentication', *Popular Music*, 21(2), pp. 209–223.

Moylan, W. (2014) *Understanding and crafting the mix: The art of recording*. 3rd edn. Oxon: CRC Press.

Moylan, W. (2020) *Recording analysis: How the record shapes the song*. New York: Focal Press.

Mudede, C. (2003) *The Turntable, CTheory*. Available at: https://journals.uvic.ca/index.php/ctheory/article/view/14561/5407 (Accessed: 2 December 2020).

Owsinksi, B. (1999) *The mixing engineer's handbook*. Vallejo, CA: MixBooks.

Owsinksi, B. (2000) *The mastering engineer's handbook*. Vallejo, CA: MixBooks.

Owsinksi, B. (2013) *The recording engineer's handbook*. 3rd edn. Boston, MA: Cengage Learning.

Pitard, J. (2019) 'Autoethnography as a phenomenological tool: Connecting the personal to the cultural', in P. Liamputtong (ed.) *Handbook of research methods in health social sciences*. Singapore: Springer, pp. 1829–1845.

Ramsey, Jr., G.P. (2003) *Race music: Black cultures from Bebop to Hip-Hop (music of the African diaspora)*. Berkeley: University of California Press.

Rights (2021) *PRS for Music*. Available at: http://www.prsformusic.com/Pages/Rights.aspx (Accessed: 10 January 2016).

Rogers, M. (2012) 'Contextualizing theories and practices of bricolage research', *The Qualitative Report*, 17(48), pp. 1–17.

Rose, T. (1994) *Black noise: Rap music and black culture in contemporary America*. Hanover, NH: University Press of New England (Music/Culture).

Schloss, J.G. (2014) *Making beats: The art of sample-based Hip-Hop*. Middletown, CT: Wesleyan University Press (Music/Culture).

Schwandt, T.A. (2014) *The SAGE dictionary of qualitative inquiry*. Los Angeles, CA: Sage.

Scott-Hoy, K.M. (2009) 'Beautiful here: Celebrating life, alternative music, adolescence and autoethnography', in B.-L. Bartleet and C. Ellis (eds) *Music autoethnographies: Making autoethnography sing/making music personal*. Bowen Hills: Australian Academic Press, pp. 39–56.

Seay, T. (2012) 'Capturing that Philadelphia sound: A technical exploration of Sigma Sound Studios', *Journal on the Art of Record Production*, 6.

Senior, M. (2015) *Recording secrets for the small studio*. 2nd edn. Oxon: Focal Press.

Sewell, A. (2013) *A typology of sampling in hip-hop*. Unpublished PhD thesis. Indiana University.

Shocklee, H. (2004) '"How Copyright Law Changed Hip Hop". Interview with Public Enemy's Chuck D and Hank Shocklee. Interviewed by K. McLeod for Alternet.org, 1 June'. Available at: https://www.alternet.org/2004/06/how_copyright_law_changed_hip_hop/ (Accessed: 20 July 2020).

Small, C. (1998) *Musicking: The meanings of performing and listening*. Middletown, CT: Wesleyan University Press.

Stavrou, M.P. (2003) *Mixing with your mind: Closely guarded secrets of sound balance engineering*. Mosman: Flux Research.

Thompson, A. and Greenman, B. (2013) *Mo' meta blues: The world according to Questlove*. 1st edn. New York: Grand Central Publishing.

Vermeulen, T. and Van Den Akker, R. (2010) 'Notes on metamodernism', *Journal of Aesthetics & Culture*, 2(1), pp. 56–77.

West, K. (2004) *The College Dropout* [CD, Album]. UK: Roc-A-Fella Records.

Williams, J.A. (2010) *Musical borrowing in hip-hop music: Theoretical frameworks and case studies*. Unpublished PhD thesis. University of Nottingham.

Zagorski-Thomas, S. (2009) 'The medium in the message: Phonographic staging techniques that utilize the sonic characteristics of reproduction media', *Journal on the Art of Record Production*, 4.

Zagorski-Thomas, S. (2010) 'The stadium in your bedroom: Functional staging, authenticity and the audience-led aesthetic in record production', *Popular Music*, 29(2), pp. 251–266.

Zagorski-Thomas, S. (2014) *The musicology of record production*. Cambridge: Cambridge University Press.

Zahavi, D. (2021) 'Applied phenomenology: Why it is safe to ignore the epoché', *Continental Philosophy Review*, 54(2), pp. 259–273.

Zak III, A.J. (2001) *The poetics of rock: Cutting tracks, making records*. Berkeley: University of California Press.

0

SAMPLE-BASED HIP HOP AS METAMODERN PHONOGRAPHIC PRACTICE (AN AUTOETHNOGRAPHY OF OSCILLATING BETWEEN—*AND BEYOND*—ANALOGUE NOSTALGIA AND DIGITAL FUTURISM)

At the end of my first ever academic paper presentation, sometime in the summer of 2016, the conference organiser—an old colleague of mine—asked: "you know this is impossible… why are you doing it?" I jokingly responded, "masochism", before providing a somewhat more scholarly answer. The impossibility he was referring to was the creation of sample-based Hip Hop out of self-made music samples. This was the objective I had set out to—practically—pursue, and—theoretically—question, as part of a doctoral research project that paved the foundations for this book; itself the next step in a ten-year career as a hip-hop artist, and incorporating my fourth solo album as applied context.

During the presentation, I played a snippet of a blues idea I had performed and recorded in the home I had been renting, for the purpose of turning it into a hip-hop sample. It featured upright piano, blues harp, acoustic drums, and electric bass. Having occupied the first floor of that house during the making of my last three albums, the place had been gradually converted into a home studio; not unlike the DIY setup Joe Meek had fashioned out of the rented property depicted in the *Telstar* documentary (*Telstar: The Joe Meek Story* [DVD], 2008), albeit featuring somewhat less vintage equipment. The house's box room—functioning as a drum booth—had been treated with re-upholstered acoustic material deemed obsolete by the university where I used to work at as a music production lecturer. I had turned one of the bedrooms into a control room, with cables connecting its mixing console to the adjacent rooms' microphones. The microphone cables could just about fit under the old house's skewed doors (and in one case, I had to saw off the lower tip of a door to squeeze through a thicker cable loom, hoping the landlord would not notice). Thick rugs doubled as both absorption material and means to cover the cabling. In what was the actual bedroom, I sacrificed the space previously occupied by a double bed to fit in the piano, and now a sofa bed situated across it provided mild absorption and a seating/sleeping solution. Inspired by my trips to maverick studios in the US—particularly Chess Records in Chicago and Sun Studio in Memphis—I would sometimes use the bathroom as an 'echo chamber', connecting a speaker and microphone to my recording interface, sending the instrumental recordings to the bathroom speaker facing the tiles, and experimenting with

DOI: 10.4324/9781003027430-2

capturing different levels of reflection at varying angles.[1] I had definitely been emboldened by researching the 1950s "sound of musical democracy" (Zak III, 2018).[2]

Comparing the blues examples *before* and *after* the 'echo chamber' treatment, I remember scanning the conference floor for reactions—as I would often do when rapping in concerts—and noticing the co-organiser, also a colleague, meta-bob positively to the echo-chamber-treated ride cymbal, extending what I perceived as a sonically satisfied smile. I appreciated the silent feedback, as I felt rather self-conscious about only being able to play the piano well—my main instrument, long before I fronted any of my beats as MC. I had only picked up the bass a few years prior, and this recording featured some of my first-ever attempts at performing acoustic drums and the blues harp. But I kept reminding myself that the point here had been to create and then capture useful phonographic moments for subsequent sampling, not to showcase musical 'chops'. Furthermore, these moments had to be steeped in vintage sonic signifiers—"sonic signatures" (Zagorski-Thomas, 2014, pp. 66–69)—as far as the recording and mixing artefacts were concerned, in order to be effective; or so my initial hypothesis went. I was trying to reach beyond the more abstract music domain arguments dominating hip-hop musicology and demonstrate the implications of the sonic materiality (the architectural make-up) of the phonographic moment for sample-based music producers—what Zak (2001, p. 89) refers to as phonographic "ephemera". My aim had been to analyse the sonic mechanics and underlying patterns of what makes an effective, impactful, and—dare I say—'authentic' sample. Zak (2001, pp. 41–42) highlights the impact that a recognition of sonic materiality can have on musicological analysis:

> What must not be overlooked, however, is that records, unlike scores, also have *material* content. That is,... they insist as well on being exactly what they are: sound, directly experienced. Interpretation that fails to take this into account will inevitably distort the picture in some way.
>
> *(original emphasis)*

The research project had, thus, been originally entitled: "Applying vintage production techniques to contemporary Hip Hop in pursuit of 'sample-based' impact and authenticity: Producing multiple records within a record". But in its revised and current title, I have only retained a paraphrasing of the latter part, which encapsulates the praxis ('Reimagining sample-based Hip Hop: Making records within records'). I gradually came to view the sonic manifestation of the *past* in samples—via the deployment of vintage production techniques—as more of a surface (textural), albeit important, characteristic (this is extrapolated in detail in Chapter 3). But where, and why, did this painstaking journey towards the creation of original sample content commence?[3] And how does this interplay between (re)engineering analogue production signatures and contemporary digital praxis lead to metamodernism?

Wayne Marshall (2006, p. 869) has accurately identified the predicament facing contemporary beat-makers in his article, 'Giving up hip-hop's firstborn: A quest for the real after the death of sampling':

> Producers working for large record labels, enjoy production budgets that permit them to license any sample they like... Independent and largely local artists, operate well enough under the radar to evade scrutiny or harassment and continue to sample with

> impunity… Acts with a sizeable national, if not international, following but who lack the resources of a "major label"—find themselves in a tight spot: to sample or not, to be real or not, to be sued or not?

That is exactly the 'spot' I found myself in at the brink of signing a major-label contract with my second (and rather sample-laden in its pre-production) album. Although my presumptions of benefitting from major-label resources did align with those described by Marshall, the national—rather than international—exposure I was about to embark on, brought me face to face with the dilemmas of a worldwide beat-maker majority: *having to seek alternative routes towards sample-based authenticity.* It was also clear in my mind—as myself and musicians in my immediate network could indeed play instruments rather well—that this was an issue of sonic, not motivic, authenticity. The experiments of this and the next album made it clear that there was an essential difference between chopping, juxtaposing, and manipulating segments of *recordings,* as opposed to doing so with segments of *records*; and even a momentary snippet of the former rather than the latter carried notably different aesthetic value within a sample-based context. The Bomb Squad's Hank Shocklee (2004) encapsulates this issue best:

> We were forced to start using different organic instruments, but you can't really get the right kind of compression that way. A guitar sampled off a record is going to hit differently than a guitar sampled in the studio…. It's going to hit the tape harder. It's going to slap at you…. So those things change your mood, the feeling you can get off of a record. If you notice that by the early 1990s, the sound has gotten a lot softer.

In Chapter 3, I theorise about the aesthetic issues with making one's own samples, and question the function of—and need for—the manifestation of the 'sonic past' within them. I observe that sample-based Hip Hop sonically celebrates the interaction of old and new music as part of its recipe. Therefore, beat-makers who create their own source content consciously invoke vintage sonics in its production, to stylise it appropriately for, and inspire, subsequent sample-based music making. The trend can be observed in the production approaches of many contemporaries, from Portishead and Boards of Canada, through to J.U.S.T.I.C.E. League, Frank Dukes, Marco Polo, and De La Soul (for more on this, see Exarchos, 2018). To offer a number of representative examples, we can observe: De La Soul with *And The Anonymous Nobody* (2016), getting themselves out of their "digital limbo" by recording over 200 hours of live music to then sample (Cohen, 2016)[4]; J.U.S.T.I.C.E. League adopting a thoroughly researched form of sonic archaeology to power the orchestral backbone to hits such as those by Rick Ross (Law, 2016); and Frank Dukes becoming a multi-instrumentalist and sample-library company owner on his way to reverse-engineer original, but retro-sounding samples (such as those powering hits for Kanye West and others) (Whalen, 2016).

The apparent irony in such a (re)constructive notion, however, lies in the creation of original music recordings, only to imbue them with artificial retrospective qualities. I found myself struggling with this conundrum. I knew my practice of making new samples of stylistically old (and old-*sounding*), yet original, music was an honest reaction to a creative issue I had faced in my professional work; and one shared by beat-makers worldwide. As an adolescent who played keyboards in bands—and forever jammed with friends and colleagues as a means of socialising *and* communication—I also really enjoyed performing in the more traditional sense, and truly loved many different forms of music (especially the Blues, Funk, Punk, and various other forms of heavy Rock). In fact, I vividly recall how *Ill Communication* by the Beastie Boys (1994) unified all stylistic dichotomies I appreciated under one lo-fi, noisy roof and became an album that defines me to this very day; it also worked as a legitimisation of my diverse tastes to my rather stylistically 'tribal' groups of high school friends. One of the reasons I think Hip Hop did become my main expressive vehicle is the very inclusiveness of other musics within it: being *inter-stylistic* by default (I expand on the inter-stylistic aspect of Hip Hop, discussing the material implications of compositional hybridisation, in Chapter 1). So, it is no surprise that the culmination of my musical and academic careers under this project became an experiment in sampling my other stylistic personas under the very boom-bap aesthetic that defined the decades that... defined me. In the years since the project begun, I would spend periods of three-to-six months completely immersed in living, breathing, listening to, playing, and recording just one particular genre of music, which I would subsequently sample (see Figure 0.1 for a stylistic mapping of source content to end sample-based productions in the accompanying album). My process consciously resonated with Simon Frith's (1996, p. 111) dictum that: "Making music isn't a way of expressing ideas; it is a way of living them". The stylistic personas inhabited through active participation in these musical genres have enabled the construction of serial musical identities, simultaneously both ideal, "what [I] would like to be", and real, participating in the "social world[s] ... enacted in musical activities" (Frith, 1996, p. 123). Sometimes, the approach involved picking up new instruments characteristic of the genre/style and practising them until I could perform them functionally enough on record. I would filter my listening habits to only relevant discography and playlists during these periods. I would also consistently study the sonics of that era, representative studios, and labels (which often involved research trips to these locations): what were the rooms, instrument setups, amplifiers, microphones, mixing desks, signal paths, and production workflows deployed?

And, yet, a number of underlying issues made me uneasy. Firstly, was this synthesis inorganic and led by a *script*? And by script, I refer both to the hypothesis at the heart of this project, *and* the creative frame drawn by the sample-licensing limitations highlighted by Marshall. I took solace in the fact that expert peers whom I respect—such as the artists/producers mentioned above—were also indulging in the self-sampling practice, and this kept me going creatively.[5] Secondly, I was aware that my classical piano upbringing in the WAM tradition had inadvertently planted some lone-white-man-genius seeds in my inherent respect barometer (and compositional ideology).[6] The classical music paradigm (and especially the pedagogy perpetuating it) had come to conflict with my 'real' musical life before, first when I joined high school bands and found that any dexterity and sight-reading skills I may have acquired from it were pretty useless without a basic grasp of the blues scale, or a modicum of improvisational ability.[7] In the context of falling in love with Hip Hop—and much other sample-based Electronica of the 1990s, from The Prodigy through to Massive Attack and Fat Boy Slim—the conflict manifested again when I realised that the sonic *aesthetics* I was drawn to were, in fact, largely sample-based in their *poetics*. Sitting in my first computer-sequencing class of a Music Technology degree in 1996, I was aware that my Electronica sounded inferior in comparison to my influences, courtesy of its dependency on legitimate yet sterile sound libraries blindly adhering to the General MIDI protocol. Furthermore, the music (and sonics) I was drawn to resonated in no small way with Afrological, cyclic sensibilities as eloquently explained to the academic world by Rose (1994) in *Black noise*,

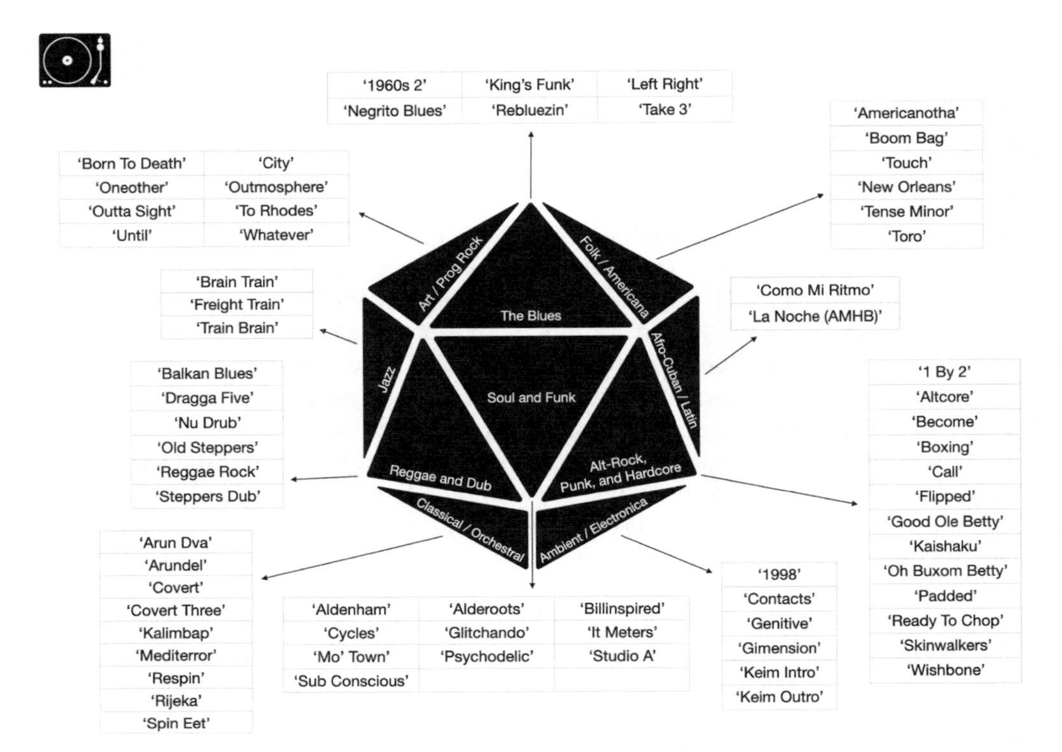

FIGURE 0.1 A mapping of the genres/styles of source audio produced to facilitate the beat-making process, mapped against tables of the final sample-based productions comprising the accompanying album. Note that the mapping only indicates source content that serves a foundational function in the end beats—there is much overlap between source elements from different stylistic categories and secondary layers in many of these productions (for example, hardcore vocal samples serving an ornamental function over classical/orchestral fundamental layers in beats like 'Arundel', or Augustus Pablo-inspired melodica from the dub pool of source content serving as a lead element in soul/funk-driven beats like 'Billinspired'. Drum samples tend to be the most featured element behind such overlaps—percussive rock or soul loops can be found behind virtually every end beat).

Ramsey (2003) in *Race music,* or Loren Kajikawa (2015) in *Sounding race in rap songs* (and other excellent scholars; in reference to Jazz and improvisation, for example, see Lewis, 2017). To sum up, I was questioning whether the synthesis this project was attempting was: a) (pun intended) too synthetic (read scripted); and b) too much of a compromise (or resolution?), opening the door for WAM compositional objectives to trickle in to the Hip Hop I was trying to make (even if this was a case of legislative necessity becoming the mother of creative invention). Although initial experiments filled me with hope, philosophically, I was still worried about the creative legitimacy of the journey rather than the actual outputs at this stage. I guess my colleague's words still resonated: "you know this is impossible...". Was this a valid way of making music? How did I fit in within the wider fabric of contemporary "musicking" (Small, 1998), politics, culture, and art?

Enter Metamodernism. Much like Shia LaBeouf's eureka moment (*The Philosophy of SHIA LABEOUF – Wisecrack Edition*, 2016), or Abramson's (2018) discovery (beautifully and reflexively told in his blog entry 'On metamodernism'), when I came across Vermeulen and Van Den Akker's (2010) 'Notes on metamodernism', and Turner's (2011) 'Metamodern manifesto', I felt the tectonic plates shift underneath my musical and philosophical feet. I felt understood. I felt explained. After further reading, I realised that the tectonic plates in question had actually shifted much earlier in the new millennium—the alignment I was experiencing was simply a delayed recognition of my own condition as part of this new "structure of feeling" (Vermeulen and Van Den Akker, 2010).[8] And my convoluted practice started to make sense as part of it, too; it was a "romantic response to crisis"—a phrase consistently used to describe much metamodern art today (Abramson, 2017b).

But let me rewind for a moment... What *is* Metamodernism? Proponents of Metamodernism describe it is as "a new cultural, political, scientific, and social movement representing a post-ideological, open source, globally responsive, paradox resolving, grand narrative" (Cooper, 2017); or more laconically, a "cultural philosophy of the digital age" dominant since about 2005 (Abramson, 2017b), "meant to replace postmodernism [which] can't come soon enough" (Cooper, 2017); also "a philosophical system, an intellectual stance, an artistic sensibility, the current cultural zeitgeist, or the 'structure of feeling' of the times in which we live" (Salguero, 2019). Its mantra? That "reconstruction must (finally) follow deconstruction" (Freinacht, 2015).

> Generally speaking, metamodernism reconstructs things by joining their opposing elements in an entirely new configuration rather than seeing those elements as being in competition with one another. If postmodernism favored deconstructing wholes and then putting the resulting parts in zero-sum conflict with one another—a process generally referred to as "dialectics"—metamodernism focuses instead on dialogue, collaboration, simultaneity, and "generative paradox" (this last being the idea that combining things which seem impossible to combine is an act of meaningful creation, not anarchic destruction). Metamodernists will often say that they "oscillate" between extremes, which really just means that they move *so quickly* between two extremes that the way they act incorporates both these two extremes *and* everything between them. The result is something totally new.
>
> *(Abramson, 2017b, original emphasis)*

The oscillation described hit a nerve when it came to how I felt transitioning between my blues, funk, rock, country, or punk personas; but also my roles as their recording, mixing, or even mastering engineer; and finally as the beat-maker juxtaposing and manipulating these previous creations. In a recent article where I was examining the pedagogic implications of the idea, I noted:

> Sample-based Hip-Hop is a form of 'meta' record-production process, as it involves the application of phonographic processes upon material that has itself been the result of a phonographic process ... Part of its mechanics is this very manipulation of content that was created without the meta-genre in mind: a funk or soul record made for its own sake, with its inherent syncopation used or abused, exaggerated and overexposed through repetition and reprogramming, chopping and truncating within the new

> context. This raises important questions about the amount of distance that should be practised when creating source content for incorporation into a sample-based hip-hop approach.
>
> *(Exarchos, 2018, p. 55)*

The research journal maintained in parallel to this book is full of reflections about getting too attached to 'songs' made in the earlier phases, initially creating beats that were far too respectful of the original structures, and eventually accumulating enough source material to afford sufficient distance from the raw content.[9] I found out that this "overlap (of) multiple subjectivities – … allowing yourself to be many different people at once without putting any one of them at the forefront" (Abramson, 2017b) was also congruent with the metamodern condition; importantly, and conversely to postmodernism, "metamodernism holds that overlapping different identities [does not destroy but] only *empowers* all of them" (Abramson, 2017b, original emphasis). As I continued studying metamodernism as an artistic sensibility, I realised that its proponents had become rather systematic in identifying traits that make one metamodern, providing detailed manifestos, colourful descriptions, typologies of principles, and even comparative charts against modernism and postmodernism. Cross-referencing the discourse against the nuances of the self-sample-based practice, I could not ignore the correlation, indicating that the process had to be an artistic manifestation of the larger paradigm.

Later beat-making experiments evolved to involve a re-composition of the constructed source materials' staging architecture, by allowing myself to open the door to multitrack elements rather than just the end (stereo) master (see Chapter 6); this was more akin to remixing really. Initially, I felt the access limitation was more 'authentic' processually, but an interview with Amerigo Gazaway (2016)—shedding valuable light on his multitrack-sampling mashup approach—and my new-found metamodern boldness, empowered me to explore the rupture.[10] As Abramson explains:

> A great example of a metamodern phenomenon is the "remix." When a musician remixes someone else's work, in a sense what they've created is dependent upon what someone *else* has created—that's one "pole"—and in a sense what they've made is something entirely *new* (a second pole). Is a "remix artist" a scavenger or a creator? Is their artwork old or new? Well, both! And therefore, in a sense, neither. In other words, when you're simultaneously creating and scavenging you are doing something that is both of those things and *also* an entirely new thing. Namely, you're *remixing.*
>
> This idea of "doing both of two very different things at once to create something new" leads many metamodernists to use the shorthand phrase "'both/and' thinking."
>
> "Both/and" means thinking that's "both" of two things "and" (therefore) something entirely new.
>
> *(Abramson, 2017b, original emphasis)*

It was at this point that the practice came to its most innovative. The source material was not someone else's work, but it might as well have been. It was another (one of my) persona's work. Being able to access multitrack elements rather than simply the whole master—whilst respecting their staging architectures arrived at as part of the source's mix context—blurred the lines between hip-hop sampling and remixing. It became both/and. Of course, the choice was always there, should I want to opt for limitation (and I made many beats exploring both/and approaches within them).

In Chapter 3, I propose that if sample-based Hip Hop was deemed postmodern by scholars (Potter, 1995; Reynolds, 2011), this hybrid of reconstructive sample-based Hip Hop had got to be metamodern. And, perhaps, sample-based Hip Hop has always been reconstructive in its essence: it was hardly ever satisfied with the "zero-sum" game.[11] An effective sample-based hip-hop production truly exemplifies "the metamodern awesome (...the twenty-first century equivalent of 'the sublime')"—"the feeling of understanding something, or at least thinking of it as 'coherent,' *without* being able to deconstruct it into its parts" (Abramson, 2017b, original emphasis). My version of this is discussed in Chapter 4 under the notion of 'sample magic'.

I wonder if King Tubby and Lee Perry can be regarded as the first metamodern musicians/remixers.[12] As Cooper (2019) states: "The point is that black culture was metamodern before some industrious white people rediscovered metamodernism". I hope this autoethnography is a metamodern act of expression as well.

> ...the purpose of this essay, much like the purpose of metamodernism itself, is merely radical transparency. Indeed, the point is to capture both the sincere and the cynical components of transparency, as transparency means revealing everything in a given "field" — not just what we're comfortable sharing.
>
> *(Abramson, 2018)*

Notes

1 The following #HipHopTimeMachine episode accounts some highlights from these trips and demonstrates how the echo chamber concepts were applied in a DIY sense—see Video 0.1 available from the 'Book > Videos' menu tab at www.stereo-mike.com.

2 In 'The death rattle of a laughing hyena: The sound of musical democracy' (2018), Zak demonstrates how the pop mainstream landscape of the 1950s was reshaped by records made by maverick producers in small or home studios, contributing to a precursor of what would later be defined as a 'lo-fi' or DIY sound.

3 Episode 0 of the #HipHopTimeMachine research vlog provides an autoethnographic account of the birth and aims of the project in the form of a video narrative—see Video 0.2 available from the 'Book > Videos' menu tab at www.stereo-mike.com.

4 Finn Cohen (2016) describes the unavailability of De La Soul's first six albums on downloading or streaming platforms as a "Legacy ... Trapped in Digital Limbo"; and goes on to trace the "complication ... in the language of the agreements drafted for the use of all those [phonographic] samples", which "do not account for formats other than CDs, vinyl LPs and cassettes".

5 For instance, Griselda/Shady Records collaborators Beat Butcha, Conductor Williams, Just Blaze, Alchemist, and Daringer exemplify the contemporary boom-bap resurgence in their output often via non-phonographic sampling. Westside Gunn's *Who Made The Sunshine* (2020b) is a case in point, as the artist proudly announced: "I wanna thank @daringer @beatbutcha_soi @conductorwilliams @justblaze @alanthechemist for the production of the album IT HAS NO SAMPLES!!!!!!!!" (Gunn, 2020a).

6 The WAM acronym refers to Western Art Music.

7 In her article, 'Getting it right: Why classical music's "pedagogy of correction" is a barrier to equity', Anna Bull (2021) questions "[w]here ... the balance lie[s] between getting it right and letting pupils go their own way", and asserts that in classical music this pursuit, "with its ideals of being faithful to the score and the composer's intentions ... perhaps reaches its zenith".

8 Abramson's (2016, 2017a, 2017b) Huffington posts are excellent, as are Brent Cooper's (2017, 2019) *Medium* essays on Metamodernism, and almost every article on Metamoderna.org (*Metamoderna*,

2021), alongside fictional philosopher Hanzi Freinacht's (2017) *Listening society: A metamodern guide to politics*.

9 Numerous 'vignettes' throughout the forthcoming chapters will illustrate.

10 See Chapter 1 for more.

11 I am aware of the irony that some of the practices cited in these sections (for example, remixing and sample-based Hip Hop) are framed as metamodernist, albeit originating in an earlier (in this case, postmodern) period. Their 'both/and' syntheses, however, exemplify that the oscillation between—and beyond—previous thought structures (i.e. modernism and postmodernism) is very much a characteristic of the new cultural developments articulated, superseding postmodernism. As such, metamodernism is not presented here as a prescriptive phenomenon, but rather a descriptive grouping of evolving thought patterns and artistic tendencies (such as the 'reconstructive' in sample-*creating*-based Hip Hop).

12 Of course, pre-existing repertoire has been continually reworked in all of the world's musical cultures (such as, in Indian, Persian, Irish/Celtic, or Blues musics); but the relevance of the dub pioneers cited here is that *their* form of re-construction involved 'play' with multitrack sonic objects of material dimensions (not in an abstract musical sense), morphing one popular music form into another.

Bibliodiscography

Abramson, S. (2016) *Five More Basic Principles of Metamodernism, HuffPost*. Available at: https://www.huffpost.com/entry/five-more-basic-principle_b_7269446 (Accessed: 23 August 2021).

Abramson, S. (2017a) *Ten Basic Principles of Metamodernism, HuffPost*. Available at: https://www.huffpost.com/entry/ten-key-principles-in-met_b_7143202 (Accessed: 23 August 2021).

Abramson, S. (2017b) *What Is Metamodernism?, HuffPost*. Available at: https://www.huffpost.com/entry/what-is-metamodernism_b_586e7075e4b0a5e600a788cd (Accessed: 23 August 2021).

Abramson, S. (2018) 'On Metamodernism', *Medium*, 16 April. Available at: https://medium.com/@Seth_Abramson/on-metamodernism-926fdc55bd6a (Accessed: 23 August 2021).

Beastie Boys (1994) *Ill Communication* [CD, Album]. Netherlands: Capitol Records, Grand Royal.

Bull, A. (2021) 'Getting it right: Why classical music's "pedagogy of correction" is a barrier to equity', *Music Educator's Journal*, 108(3), pp. 65–66.

Cohen, F. (2016) 'De La Soul's Legacy Is Trapped in Digital Limbo', *The New York Times*, 9 August. Available at: https://www.nytimes.com/2016/08/14/arts/music/de-la-soul-digital-albums.html (Accessed: 1 February 2017).

Cooper, B. (2017) '"Beyond" Metamodernism: The Meta- Turn has Come Full Circle', *Medium*, 16 April. Available at: https://medium.com/the-abs-tract-organization/beyond-metamodernism-c595c6f35379 (Accessed: 23 August 2021).

Cooper, B. (2019) 'Black Metamodernism: The Metapolitics of Economic Justice and Racial Equality', *Medium*, 1 June. Available at: https://medium.com/the-abs-tract-organization/black-metamodernism-a72d24da6f0f (Accessed: 23 August 2021).

De La Soul (2016) *And the Anonymous Nobody* [CD, Album]. US: AOI Records.

Exarchos, M. (2018) 'Hip-Hop pedagogy as production practice: Reverse-engineering the sample-based aesthetic', *Journal of Popular Music Education*, 2(1&2), pp. 45–63.

Freinacht, H. (2015) *You're Not Metamodern before You Understand This. Part 2: Proto-Synthesis, Metamoderna*. Available at: https://metamoderna.org/youre-not-metamodern-before-you-understand-this-part-2-proto-synthesis-2/ (Accessed: 23 August 2021).

Freinacht, H. (2017) *The listening society: A metamodern guide to politics*. Metamoderna.

Frith, S. (1996) 'Music and identity', in S. Hall and P. Du Gay (eds) *Questions of cultural identity*. Los Angeles, CA: Sage, pp. 108–128.

Gazaway, A. (2016) 'Interviewed by author, 11 August'.

Gunn, W. (2020a) *I Wanna Thank…, Instagram*. Available at: https://www.instagram.com/westside-gunn/ (Accessed: 15 September 2021).

Gunn, W. (2020b) *Who Made The Sunshine* [Digital Release, Album]. US: Shady Records.

Kajikawa, L. (2015) *Sounding race in rap songs*. Oakland: University of California Press.

Law, C. (2016) *Behind the Beat: J.U.S.T.I.C.E. League, HotNewHipHop*. Available at: https://www.hotnewhiphop.com/behind-the-beat-justice-league-news.23006.html (Accessed: 6 September 2017).

Lewis, G.E. (2017) 'Improvised music after 1950: Afrological and eurological perspectives', in C. Cox and D. Warner (eds) *Audio culture: Readings in modern music*. 2nd edn. New York: Bloomsbury Academic, pp. 385–398.

Marshall, W. (2006) 'Giving up hip-hop's firstborn: A quest for the real after the death of sampling', *Callaloo*, 29(3), pp. 868–892.

Metamoderna (2021). Available at: https://metamoderna.org/ (Accessed: 23 August 2021).

Potter, R.A. (1995) *Spectacular vernaculars: Hip-hop and the politics of postmodernism*. Albany: State University of New York Press.

Ramsey, Jr., G.P. (2003) *Race music: Black cultures from Bebop to Hip-Hop (music of the African diaspora)*. Berkeley: University of California Press.

Reynolds, S. (2011) *Retromania: Pop culture's addiction to its own past*. New York: Faber and Faber.

Rose, T. (1994) *Black noise: Rap music and black culture in contemporary America*. Hanover, NH: University Press of New England (Music/Culture).

Salguero, P. (2019) 'My Initial Reactions to Metamodernism', *Medium*, 9 September. Available at: https://piercesalguero.medium.com/notes-on-metamodernism-3ad31b450886 (Accessed: 23 August 2021).

Shocklee, H. (2004) '"How Copyright Law Changed Hip Hop". Interview with Public Enemy's Chuck D and Hank Shocklee. Interviewed by K. McLeod for Alternet.org, 1 June'. Available at: https://www.alternet.org/2004/06/how_copyright_law_changed_hip_hop/ (Accessed: 20 July 2020).

Small, C. (1998) *Musicking: The meanings of performing and listening*. Middletown, CT: Wesleyan University Press.

Telstar: The Joe Meek Story [DVD] (2008). G2 Pictures.

The Philosophy of SHIA LABEOUF – Wisecrack Edition (2016). Available at: https://youtu.be/6dsEC-bVahBw (Accessed: 19 July 2020).

Turner, L. (2011) *The Metamodernist Manifesto | Luke Turner (2011), The Metamodernist Manifesto | Luke Turner (2011)*. Available at: http://www.metamodernism.org (Accessed: 23 August 2021).

Vermeulen, T. and Van Den Akker, R. (2010) 'Notes on metamodernism', *Journal of Aesthetics & Culture*, 2(1), pp. 56–77.

Whalen, E. (2016) *Frank Dukes Is Low-Key Producing Everyone Right Now, the FADER*. Available at: https://www.thefader.com/2016/02/04/frank-dukes-producer-interview (Accessed: 6 September 2017).

Zagorski-Thomas, S. (2014) *The musicology of record production*. Cambridge: Cambridge University Press.

Zak III, A.J. (2001) *The poetics of rock: Cutting tracks, making records*. Berkeley: University of California Press.

Zak III, A.J. (2018) 'The death of a laughing hyena: The sound of musical democracy', in R. Fink, M. Latour, and Z. Wallmark (eds) *The relentless pursuit of tone: Timbre in popular music*. New York: Oxford University Press, pp. 300–322.

PART 1

(Inter-stylistic) Composition, and tools

1

SONIC NECESSITY, MOTHER OF COMPOSITIONAL INVENTION (MAKING BLUES FOR SAMPLE-BASED HIP HOP)

Where does one start compositionally, attempting to create original source content that will serve a sample-based music-making journey? What styles of music may be fruitful, and how should they be constructed (recorded and produced) phonographically to serve this pursuit? Are there inter-stylistic tensions, as well as opportunities, situated between past genres and Hip Hop that can be explored to inform their potential synthesis? And what are the sonic and musical priorities that should be considered in this interplay? These are the key questions driving both the theoretical and music production pursuits in this chapter, examining the Blues as a past-style case in point in the context of its interaction with sample-based music making, and specifically Hip Hop.

Why start with the Blues? First, Hip Hop and the Blues share important historical and cultural associations. Second, blues samples have been exploited considerably less in Hip Hop when compared to other musical styles such as Funk and Soul. Therein lies an opportunity to investigate the cross-genre rationale for this causality, and put under the microscope contemporary music productions that exemplify effective resolutions. Finally, there is a personal reason for exploring this particular cross-genre interaction: a deep love for the Blues, and a life-long engagement with this music as a player (particularly on piano and keyboards). It is important to therefore acknowledge that for each practitioner any such inter-stylistic experiment may take different forms and explore different genres, leveraging unique synchronicities of familiarity, instrumental ability, and even cultural 'proximity'. Yet, the resolutions from this particular flux should extend beyond the genres on hand and provide conceptual food for thought (and praxis) when it comes to the inclusion of further styles within sample-based music making. Drawing from successful case studies of congruent merging offers a framework that can inform such inter-stylistic experiments, at the same time functioning as a lifeline for a practice that may be running low on renewable raw materials (i.e. phonographic samples). In the words of hip-hop producer Domino (cited in Schloss, 2014, p. 164):

> I just think that, now, you're getting to the point where … you're running out of things to find. And so a lot of the best loops have been used already. I mean, there's

DOI: 10.4324/9781003027430-4

> some stuff out there, I'm sure. There always will be stuff. But now it's like, in order to stop recycling things, you gotta just take pieces and make 'em into a whole new thing.

In the following sections, I will first discuss some of the background and historical/cultural associations between Hip Hop and the Blues, then proceed to break down the sonic and musical variables of three illustrative case studies by other practitioners. Eventually, I'll put some of the acquired 'knowledge' to the sample-based test: producing original material steeped in past sonic signifiers to create source content for beats included in the accompanying album.

Background

> It's no doubt that there's a connection [between the blues and hip-hop]. Hip-hop is definitely a child of the blues. And I think you gotta know the roots to really grow. It's [like] knowing your parents, it's like knowing your culture, so you could be proud of that culture and take it to the world.
>
> *(Common cited in Levin, 2004, p. 187)*

Recorded music history provides ample evidence of a close relationship between Hip Hop and the Blues. As early as 1930, Memphis Minnie celebrated a hit in the form of an innovative talking-blues record with 'Frankie Jean' (in Various, 1996). Just over two decades ago, rapper Nas reached out to his father—jazz-blues musician Olu Dara—on 'Bridging The Gap' (from *Street's Disciple*, 2004). And more recently, Abdominal and The Obliques produced a blues-hop album, *Sitting Music* (2012), while Amerigo Gazaway mashed up Blues and Southern Rap in the conceptual collaboration of *B.B. & The Underground Kingz* (2015). Since the birth of Hip Hop, rappers have been proclaiming their affiliation with the Blues in their statements, lyrics, and music, and scholars have drawn parallels between the socio-political backgrounds and narrative approaches of the two genres (for example: Rose, 1994; Guralnick, Santelli and George-Warren, 2004; Chang, 2007). In *Can't stop won't stop,* Jeff Chang (2007, p. 13) observes that "if blues culture had developed under the conditions of oppressive, forced labor, hip-hop culture would arise from the conditions of no work".

When Chuck D of Public Enemy discovered Muddy Waters's *Electric Mud* (1968) album, not only did he identify with blues music, but he also resonated with the experimental and dense layering of a later blues recording:

> I was sparked about the blues as a beat digger coming across an album of immense layers and well-played sounds ... Myself and my co-producer Gary G-Whiz fell in love with the record, a psychedelic trip replaying and singing Muddy's classics of the past.
>
> *(Chuck, 2003, p. 281)*

Mirroring Public Enemy's own heavily layered production style, *Electric Mud* provided Chuck D (2003, p. 280) with a rap musician's sonic window onto the past: "being a so-called veteran of the genre labeled hip-hop and rap music, you can't help being a musicologist, or at least a student of music, by default".

Yet despite many of the thematic and cultural similarities celebrated, the reality of the musical crossover between rap music and the Blues reveals a number of tensions. One could argue that hip-hop music is by default inter-stylistic, and since its very inception, it has depended on phonographic segments from other musical styles in order to function and exist. DJ Kool Herc carried over the performative sound-system tradition of using extended instrumentals on turntables from Jamaica to New York, replacing the reggae dubs with funk breaks and, thus, providing a rhythmical foundation for MCs to rap over (see, for example: Toop, 2000; George, 2005; Chang, 2007; Kulkarni, 2015; Serrano, Torres and Ice-T, 2015).[1] The practice elevated the funk drum break to *building block of choice* for future hip-hop productions, and Funk—in its various guises—remained a referential mainstay throughout all eras of hip-hop composition. James Brown's *Funky Drummer* (1970) became one of the most sampled songs in popular music history, powering the majority of East Coast's boom-bap productions;[2] while P-funk inspired West Coast's synthesiser-driven divergence via interpolation and live performance.[3] As a result, the lion's share of rap releases became literally powered by late 1960s and 1970s Funk and Soul, either by way of direct phonographic sampling or through compositional referencing;[4] and although blues samples indeed feature in Hip Hop, it will be important to question their lesser presence when compared to funk and soul sampling, despite Rap's otherwise celebrated affiliation to the Blues.

In response to the growing practices of interpolation, original composition, and live performance within hip-hop production, it will also be important to consider issues of stylistic authenticity and sonic impact arising from these alternative approaches. Marshall (2006, p. 880) discusses this dynamic in the work of live hip-hop band The Roots:

> ...the degree to which the Roots' music indexes hip-hop's sample-based aesthetic serves as a crucial determinant of the group's "realness" to many listeners. At the same time, the Roots' instrumental facility affords them a certain flexibility and freedom and allows them to advance a unique, if markedly experimental, voice within the creative constraints of "traditional" hip-hop's somewhat conservative conventions.

On the other hand, in the cases where sampling practices do have the potential to interact with original composition and performance in a synchronous (or near-synchronous) context, it may be fruitful to consider them as active determinants in the shaping of this material, as opposed to mere agents that enable the manipulation of a 'passive' recorded past. The effect of studio practices on the evolution of musical aesthetics has precedents that date back to the very beginning of phonography and, in the case of the Blues, Robert Johnson analyst Eric W. Rothenbuhler (2006, p. 78) has highlighted that: Johnson's "music reflected a then nascent recording culture ... [which] was influenced by recorded music and showed signs of being composed and performed with attention to a kind of for-the-record aesthetic". It is fair to extrapolate that when a studio process enables the creation of musical content designed to feed sample-based composition, this may be described as a case of pursuing a kind of *meta*-record aesthetics: where the meta-genre (Hip Hop) not only digests, but *shapes* the source-genre. The degree, dynamic, and potential outputs of this interaction become the subject of reflexive analysis in this chapter, extrapolating further on the effect of sampling technologies on inter-stylistic synthesis, morphing, and the creation of cross-genres. As Zak (2001, p. 73) states, "[i]n the development of a music so stylistically dependent upon machinery, the history of technology and the history of musical style are linked".

Samplin' and tumblin' (case studies)

The case studies selected here illustrate a range of approaches—spanning from interpolation, through to live performance, and remixing—offering opportunities for analysis of the musical and sonic qualities of blues recordings from the perspective of their sample 'appropriateness'. The premise being that if sampling is the predominant method in beat-making, it is important to look at (i.e. aurally analyse) representative blues works that feature within rap songs as source material. In 'Records that play: The present past in sampling practice', Vanesa Chang (2009, p. 147) explains:

> The successful pursuit of new samples has, as its limit, the producer's capacity to hear musical possibility in a song, to listen for connections that may not currently exist in the song, to perceive aural spaces where they might not be obvious. This requires conceiving of sound as plastic material, and not as a finished product.

The three cases I have selected here are: Nas's 'Bridging The Gap' (from *Street's Disciple*, 2004) referencing Muddy Waters's 'Manish Boy' (1955); Amerigo Gazaway's mashup 'The Trill Is Gone' (2015), sampling B.B. King's 'The Thrill Is Gone' (1969); and Abdominal and The Obliques' track 'Broken', from their album *Sitting Music* (2012).

Case 1: 'Bridging The Gap'

'Bridging The Gap' is a collaboration between rapper Nas and his father Olu Dara, a jazz-blues musician who performs lead guitar, trumpet, and harmonica on the track. Producer Salaam Remi performs bass, guitar, and drums, and session musician Vincent Henry is credited with the remainder of the live performances, namely baritone sax, harmonica, and strings. Although 'Mannish Boy' receives no sampling or interpolation credit—and only father, son, and producer are credited with writing and composition—the central guitar and harmonica motif can rather clearly be identified as a faster (and melodically sparser) homage to Muddy Waters's and Junior Wells's interaction on 'Mannish Boy', which is further accentuated by the melodic similarities in Olu Dara's chorus. Olu Dara's lyrics, however, are different to Muddy's version and this, perhaps, legitimises Dara's reclaiming of the motif. The adaptive approach is consistent with the early blues tradition of shared motifs and a more inclusive notion of composition, which Rothenbuhler (2006, p. 71) describes as follows:

> In the early blues tradition, as in most oral cultures, there was little emphasis on composition as we define it and value it today. Both lyrics and music were combinations of standard figures and phrases, a given performer's own adaptations or inventions, and new phrases invented or chosen from the stock to fit the situation of performance.

Ironically, this compositional position has a lot in common with Hip Hop's sampling philosophy and production ethics. Yet the adapted introductory motif here retains its blues-derived triplet feel (12/8), which cannot quite *bridge the gap* with Hip Hop's funk-derived reliance on common time (4/4). Salaam Remi does not attempt to resolve the tension, instead structuring the production around a clearly defined 'duality' of 12/8 choruses (featuring sung parts by Olu Dara) and 4/4 verses (featuring Nas's raps). The 12/8 blues hook that introduces the song is suddenly sped up and re-appropriated in common time at 0:33, punctuated by Nas's "let's go" shout initiating the verse figure, and resembling a sample-based gesture, which—although highly swung in its relationship to the syncopated drum part—nevertheless remains in 4/4. The verses are constructed around a two-bar repetition of the live drums, and the guitar and harmonica riff, with occasional solo harmonica flourishes, sixteenth snare drum fills, and strings that build up at the end of four- or eight-bar sequences. Despite the construction of the verses out of live performances, the main verse 'loop' here conveys a sample-based approach, whether the drum pattern and riff repetition are in fact constructed with the use of a sampler or looped around within a Digital Audio Workstation (DAW). The up-front placement of the drum mix, its consistent two-bar repetition, the rhythmical interruptions of the beat, and the tight placement of what feels like a 'chopped' version of the blues motif against it, convey a clear sample-based sensibility. Furthermore, the recording and mixing sonics imprinted upon the blues performances are reminiscent of vintage production qualities (such as lower fidelity and higher tube saturation, similar to mid-to-late 1950s Chess label recordings), which distances them from the more modern sonic signatures imprinted upon the drums and raps. The fact that multiple studios have been deployed for the completion of this track, may suggest that the producer purposefully pursued particular era-invoking timbres from alternate technical setups when dealing with the different instrumental groups.[5] The drum figure is reminiscent—both in its accents and sonics—of 1970s funk break-beats, such as Clyde Stubblefield's drum break from *Funky Drummer* (Brown, 1970). The sonic differentiation is further exemplified by the different timbral qualities and spatial treatments on both Nas's and Olu Dara's voices, the former appearing more contemporary and congruent with a post-2000 rap aesthetic, the latter signposting towards a more distant—if somewhat generic—past. As such, 'Bridging The Gap' highlights musical and timbral tensions between the Blues and Hip Hop, presenting the producer with rhythmical and sonic ultimata. Although the Blues are hereby 're-constructed' rather than phonographically sampled, Salaam Remi chooses to amplify the stylistic differences by dialling in structural and timbral polarities, resolving to a historically intermediate style—Funk—for his drum break, which acts as a catalyst in *bridging the gap*.

Case 2: 'The Trill Is Gone'

Amerigo Gazaway is a Nashville-based producer who is well known as a "chemist" (Caldwell, 2015) of the mashup creating "collaborations that never were" (Reiff, 2015). Having previously mixed Marvin Gaye's soul vocals with Mos Def's raps, and The Pharcyde's—*West Coast*—rhymes against Tribe Called Quest's—*East Coast*—instrumentals (Roberts, 2012), his work is identifiably sample-based. But he elevates the 'mashup' beyond its historical definition as a mere juxtaposition of two or more synchronised records. Through extensive sampling of smaller segments from multiple sources, sample manipulation, live recording, and computer programming, he is able to synthesise the numerous elements into a coherent whole of notable musicality. This is enriched by his considerable skills in live musicianship, which allow him to integrate organ, electric bass, electric piano, synthesisers, and turntables into the mix, effectively 'jamming' with the sampled musicians who *never were* in his studio. His method places him in the virtual seat of a producer who works with artistic 'ghosts' from the past, creating a metaphor of a more physical production paradigm. For *B.B. & The Underground Kingz*, Soul Mates Records (2015) state:

> Aptly titled "BB & The Underground Kingz: The Trill is Gone," the producer seamlessly bridges the gap between hip-hop and its predecessor, the blues.

> Crafting the album's bedrock from deconstructed samples of King's electric blues hits, Gazaway re-imagined what might have happened had King and UGK actually recorded in the same time and space ... Strategically looping and lacing Lucille's guitar licks and B.B.'s road tales with Bun B & Pimp C's southern fried storytelling, Gazaway finds a sweet spot in the overlapping themes of his subjects' respective catalogs.

Amerigo (Gazaway, 2016) adds: "I'm trying to get away from using that word [mashup], and trying to call it something like a conceptual collaboration". Gazaway's method may appear as a polar opposite to Salaam Remi's interpolation approach on the surface because of the precedence of sampling over 'original' composition; yet he is able to achieve a more integrated co-existence between the two genres, moving away from distinct structural dualities or triplet-based time signatures forced into common time. This is partly due to his micro-sampling processes, but also because of characteristics inherent in the *type* and era of Blues that he chooses to sample.[6] On a track such as 'The Trill Is Gone', he chooses B.B. King's 1969 version of 'The Thrill Is Gone', a 4/4 rendition of the 6/8 minor jazz-blues original released by Roy Hawkins (1951/2000) in 1951. The B.B. King version is characteristic of a late 1960s or early 1970s blues treatment as, by this point, the influence of Soul and Funk can be felt clearly on the Blues. The time signatures begin to favour common time, and many of the arrangements expand considerably to include larger sections (often brass and strings), contributing to more polished productions with larger ambient footprints and less mix saturation (particularly when compared to the mid-1950s Chicago Blues referenced for 'Bridging The Gap'). Similar characteristics can be heard on records by the other two 'King' contemporaries, Freddie King and Albert King (the latter exemplifying the soul-blues formula of Stax Records), with comparable sonic signatures on records such as *Help Me Through The Day* (1973) and *I'll Play The Blues For You, Pt. 1* (1972), respectively. The tendency for R&B-inspired, minor 4/4 Blues in this era, with spacious arrangements and extended electric guitar solos (at tempos that range between 80 and 95 bpm), is particularly helpful in the hands of sample-based music producers such as Gazaway. The link with Funk has already been established within the source material, there are no time signature tensions to be resolved, and the extended instrumental sections provide multiple opportunities for sampling particular parts. Furthermore, the fuller arrangements enrich the sampled palette with wide frequency spectra, and the minor harmony is congruent with the dark mood of much modern Hip Hop. Characteristically, on 'The Trill Is Gone', Gazaway diverts from the (funk-derived) hip-hop habit of staying on the I chord for the duration of the song (which 'Bridging the Gap' pays tribute to), and instead follows the harmonic movement of B.B. King's version. He reduces the tempo from approximately 90 to 78 bpm (and consequently the tonality of the song from Bm to Am), but respects the i-iv-i-bVI-v(7) sequence of the 1969 version. He also re-arranges various instrumental guitar segments under the rapped verses, creating a classic 'call-and-response' blues signature between the guitar and vocals. Further additions include live organ lines for the later parts of the choruses, as well as backing vocals and ad-libs.

The resulting rich and pluralistic musical arrangement is characteristic of Southern Rap's divergence from East and West Coast Hip Hop in the mid-1990s, justifying the inter-stylistic intentions here also from the perspective of Hip Hop's evolution.[7] Amerigo (Gazaway, 2016) consciously pays homage to both DJ Screw's slowed-down "chopped and screwed" sampling style and Pimp C's gospel-inspired use of live instrumentation, regarding the latter as a pioneer in "making Southern Rap music that was melodic and had harmony" and "blending the old with the new".[8] In other words, Gazaway finds commonalities between later (funk/soul-contaminated) Blues and more recent hip-hop divergences, in order to allow for richer harmonic progressions that support this complimentary meeting of genres. Through these choices he demonstrates a positive case of "trans-morphing", where he not only successfully mixes the two genres but arguably creates a new, hybrid one;[9] one that sits comfortably within the evolutionary narrative of Southern Rap. The choice to pitch and slow down the instrumental by two semitones and approximately 12 bpm supports idiosyncrasies characteristic of Southern Rap subgenres, and this may be one of the most crucial decisions Gazaway makes initially.[10] As a consequence, he accepts the reduced frequency 'presence' of the original recording's spectrum, which in turn allows him to place many of the blues samples 'behind' the programmed beat as far as the 'depth' perspective of the mix is concerned. He also abuses the two sides of the blues multitrack by widening it to an audible extent in order to allow for a distinct 'centre-stage' placement for his newly programmed electronic kick and snare drums (characteristic of the subgenre's reliance on Roland TR-808 drum-machine timbres). The original electric piano parts are exposed on the right side of the stereo image, which he chops and edits at the end of every eight-bar section to enhance their rhythmical effect. His synthetic high-hats sit comfortably on top of otherwise mildly equalised instrumental elements (another result of the pitching down process and, perhaps, his further equalisation of the samples), interplaying between eighths, sixteenths, and thirty-seconds in the high-hat programming. Finally, the expansive ambience of the original blues mix enhances the combined, illusory 'depth' effect, giving the blues signature a distinctly haunting 'space' within the architectural landscape of the mashup; it feels like past and present are occupying separate sonic spaces.

In another production decision of key importance, Amerigo (Gazaway, 2016) uses the Melodyne software to transform the original live electric bass into a MIDI part, which he then uses to trigger a "dirty south sub synth bass" at "the exact same shuffle, the exact same groove as the actual bass player that played on the record", serving the southern synth-bass sensibility but keeping "that human groove". Thus, the totality of the sonic characteristics described appears intrinsically linked to musical decisions conceived of as part of cross-genre mixing. In a move that mirrors his elegant, sonic trans-morphing, Gazaway sums up the cross-genre journey in the title of his mashup, changing *thrill* to *trill*, a term simultaneously referring to Texas slang and a Southern Rap subgenre (for more on *trill*, see Bun B's interview in Harling, 2013). Finally, he identifies his contemporaries' fear of dealing with triplet subdivisions as the main reason behind the less frequent integration of Blues and Rap, a creative challenge that he wholeheartedly accepts on other tracks of the *B.B. & The Underground Kingz* (2015) album by "working with it a little bit more, massaging it [further] and pushing it … more" (Gazaway, 2016).

Case 3: 'Broken'

At the other extreme of the blues-rap spectrum are situated attempts at a fully live-performed Hip Hop, borrowing from traditional blues composition and performance practices. Perhaps the most representative live hip-hop band are The Roots, while *Blakroc* (2009)—the collaborative album between hip-hop producer Damon Dash and rock group the Black Keys—also provides a relevant case. The Roots, however, owe more to Jazz, Funk, and Soul than to Blues directly, and although the Black Keys are often referred to as a blues-rock act, *Blakroc* mixes alternative and garage rock influences in equal measure. Abdominal and The Obliques, on the other hand, are one of the very few acts that receive the quintessential #BluesHop tag in the online world, exemplifying the cross-genre as a hybrid of live instrumental performance and rapping. The group formed as a side project of Toronto rapper Andy Bernstein—known by his stage name as Abdominal, sometimes Abs—and released their album *Sitting Music* (in 2012), offering a useful case study of hybridisation with its own compositional and sonic resolutions. Thomas Quinlan (2012) provides the following review:

> A mix of folk, blues and country [that] replaces the sampling and boom bap beats. Revitalized by a backing band – guitarist Andrew Frost and percussionist Colin Kingsmore – Abs is still rapping but with a smoother flow that sometimes becomes singing, while his band provide backing harmonies and hooks … Sitting Music might not be your typical hip-hop album – Abs describes it as blues-hop and "middle-aged hip-hop" – but it's great to see an artist striving to stretch his boundaries.

The very inception of the project is a consequence of Andy Bernstein's (2016) reaction to growing older and wanting to experiment with more introspective lyrical themes, which required a different sound and "some time apart" from what he describes as "traditional Hip Hop". Bernstein (2016) explains:

> My whole career has been more [about] doing the straight-ahead rap stuff, like typical rap-beat-samples, that kind of thing. The idea behind [forming] the band was just really because I was getting older and I was finding [that] I wanted to tackle some new themes … slightly more introspective themes, look at some vulnerable kind of topics. So, it just didn't feel right to me to have the same—the usual—typical sample-based boom-bap rap beats for those types of songs. That was really the main reason for me to put the band together, just to kind of get like a mellower, quieter sound that would better fit the themes of these new types of songs.

Abdominal (Bernstein, 2016), furthermore, disagrees with the notion of a direct link existing between rap authenticity and the sample-based method, because Hip Hop "does not equal one particular sound, it's more (of) an approach, an aesthetic … using what's around you and crafting it to form something new, whether it's using samples, whether it's, oh you know, I know this guitarist…". On single 'Broken', Bernstein, Frost, and Kingsmore use percussion, and acoustic, electric, and slide guitars, all recorded in a domestic basement with household objects used as separation baffles, and vocals overdubbed at the rapper's home studio. Abdominal's vision for the album was to capture the sonic of "just three dudes sitting on a porch playing", something they achieve by focusing on simultaneous performances as much as possible, allowing recording 'spill' to take place, and not overly polishing the post-production process (Bernstein, 2016).[11]

The general harmonic progression of I-bIII-I-IV-V in the choruses and I-bIII-ii-V in the verses (with frequent usage of passing chords and extensions), and the laid-back performing style over a slow tempo conjure a 'swamp blues' feel; but there is a definitive hip-hop influence on the hybridisation of the composition and arrangement. The percussive beat is simplified to quarter and eighth accents resembling a drum-machine pattern, with shakers added progressively to emulate programmed high-hat sixteenths. At 01:43 Frost and Kingsmore perform a quarter-note 'stutter' echoing a sample-based repeat which lasts one additional bar, while on many occasions (i.e. at 1:30, 2:50, and 3:23) there are complete instrumental stops resembling DJ 'cuts'.[12] The main acoustic guitar sequence repeats throughout the verses assuming a 'looped' function, often ending the four-bar sections with rather exposed and mechanised quarter strums, 'marking time' so to speak. Although the intention here may not be to pursue a sampled or programmed production texture, it is clear that the musicians' experience of Hip Hop, and their catering for the rapped verses, drive their compositional, arranging, and structural decisions towards an effective hybridisation: one that is thematically supported by the lyrics and the group's open-minded approach to experimentation in negotiating the two genres. Notably, they remain less experimental during the rapped verses, with a simpler harmonic progression and a strict four-bar repetition, while for the sung choruses they support the melody by leaving the tonic for the flat-third chord, then returning to the tonic before the closing IV-V-I turnaround. The resulting five-bar chorus cycle feels supportive of the sung melody, giving the hook a distinctive if peculiar feel, while parallels can be drawn to early recorded country or folk Blues where performers would extend their own accompaniments to cater for the uniqueness of their melodic or lyrical lines.[13] It is not a surprise that out of the three case studies, it is the live blues-hop production that presents the most compositional freedom, but what is important here is the influence of the *meta*-genre on the traditional form, even without its form-shaping technologies directly on hand (i.e. the use of samplers and DAWs). In effect, the sampling practices that have shaped the 'meta-genre' are not hereby utilised directly, but their stylistic aftermath is exercised by the musicians in absentia, shaping the very rhythm, arrangement, and structure of their blues-inspired performances.

Meta-jamming: setting up the inter-stylistic experiment

As part of the wider project, I have written more than eight hours of original blues content in preparation for the sampling phase, referencing blues styles from the 1950s to the 1970s. For the purposes of this particular experiment, I have referenced late-1960s to early-1970s minor blues examples, such as the southern Blues that came out of Shelter Records in Texas, and Stax Records in Memphis. The aim has been to create a relevant applied context, which will welcome practical exploration of the findings from the three case studies above: musical and sonic characteristics that reflect the first two case studies, but also compositional freedom reflected in the third. Specifically, a 15-minute improvisation has been conceptualised and then recorded by overdubbing acoustic drums, electric bass, upright piano, electric Rhodes piano, electric guitar, and shaker. Loosely inspired by the aforementioned references, the improvisation has taken place over an eight-bar iv-i-V(7)-i harmonic progression (for the verses) and a bVI(7)-i-V(7)-i variation (for the bridges or choruses), at a harmonic speed of two bars per chord, a tempo of 85 bpm, and a time signature of 4/4 (see Table 1.1).

The drums were recorded to a metronome click and the rest of the instrumentation was in synchronisation with the drums, with the aim of aiding the editing processes of the forthcoming beat-making phase.[14] The duration of the recording was extended to 15 minutes—or 642 bars—supporting the development of instrumental synergies and furnishing the sampling phase with a rich palette of options. The cyclic blues form consisted of

TABLE 1.1 Harmonic progression of the blues recording

Bars	*1*	*2*	*3*	*4*	*5*	*6*	*7*	*8*
Verse	iv	iv	i	i	V(7)	V(7)	i	i
Bridge /Chorus	bVI(7)	bVI(7)	i	i	V(7)	V(7)	i	i

pattern and dynamic variations centred around a main electric bass guitar motif often coupled with the electric guitar (a figure frequently employed at Stax by Albert King and bassist Duck Dunn of Booker T. and the M.G.'s), while the electric Rhodes piano supported the harmony and rhythm with chordal work in the middle register. The upright piano provided rhythmical and harmonic support initially, then delved into solo improvisation as the track progressed. The drums gradually developed from simple eighth bass-drum and snare-drum patterns using the cross-stick on the snare, to more syncopated and swung sixteenth accents progressively deploying the full snare. These were consciously performed to mirror a range of references, again with the aim of enriching the potential sampling pool of the later phases.[15] After minor macro-editing of the performances—which aimed at preserving the micro-level interaction between instrumental performances and their resulting 'groove', but nevertheless removing any content of no use—125 segments of half, single, and dual bars were deemed as worthy samples, each segment representing no more than a single chord in the harmonic progression. These were then exported as synchronised multitrack components and (given the track-count of 15 channels for most parts of the structure) resulted in 1,854 audio files.[16] The synchronised stereo segments were brought into another DAW for mixing, utilising software emulations of representative hardware technologies for the era. Particular attention was directed towards microphone pre-amps, mixing desk summing, and recording format (tape, vinyl) colouration, characteristic of vintage sonic signatures imprinted on material that would frequently be favoured for sampling. Master tape and vinyl record emulations of the stereo files were prepared for each segment of the multitrack, and the mixed results were exported as 24-bit wave files, compatible with Akai's MPC Renaissance music production controller, which was to be used extensively in the following sampling phase.[17] Finally, each segment or 'chop' was assigned to a drum pad on the MPC, taking up most of its eight banks of 16 pad locations (a maximum of 128 per program) to fuel the following phase.[18]

The Beastie Boys (2022) have described this long-winded process of creating source material for sampling as similar to "making a two-year-old pizza", when talking about their approach to recording *Check Your Head* (1992):

> On one particular tape was a long space[d] out jam that had a few sensational sounding bars of playing on it. We sampled those bars on Mike's MPC 60 sampling drum machine. Put them in a kind of arrangement, Yak wrote a few lyrics and we each took a turn singing those words. In a samply kind of way, we muted and unmuted the vocal tapes that sounded good. The end result is the song 'Something's Gotta Give'. It certainly is odd that a three and a half minute song would take two and a half years to finish. But that's how this record was made. Like you'd make a dough then put it on top of the fridge for a while to rise. You'd make a sauce, add ingredients, let them sit together ... you get where I'm going with this, right? I'm making a two-year-old pizza.

Video 1.1, available from the 'Book > Videos' menu tab at www.stereo-mike.com, showcases a short section of the blues improvisation recorded over a number of instrumental overdubs (with embedded captions).

Chopping the Blues: sample-based composition

A lot has been written about 'sample-based' composition, some of it polemic (for example: Goodwin, 1988) and some supportive (for example: Rodgers, 2003; Harkins, 2008, 2010; Schloss, 2014; Swiboda, 2014; Williams, 2014), while much of the literature is focused on the ethical and legal dimensions of what is regarded as new or original work (for example: McLeod, 1999; Collins, 2008). Although, as we have seen above, the Blues themselves challenge Eurocentric notions of composition, in this particular case, the publishing and mechanical constraints that would limit sample-based composition are removed by virtue of the practitioner both *sampling* and being *sampled*. This context allows for a focused reflexive analysis of the interaction between sampling practice and the construction of pre-recorded material, without diluting the question with peripheral concerns. Furthermore, it will be useful to extrapolate on potential synergies resulting from this closer relationship between the two functions: composer as content creator, as well as content 'manipulator'.

The Akai MPC range facilitates a particular sampling workflow due to its interface design, operating system, but also a number of inherent sonic characteristics (these are examined in more detail in Chapter 2). The drum pads situated 'on top' of its interface—for all of its hardware, software, or hybrid incarnations—invite a percussive style of triggering musical material, while the Roger Linn-derived rhythmic quantisation (with its characteristic swing and inherent timing imperfections) is the subject of much reverence from scholars and practitioners alike (see, for example: Rose, 1994; Schloss, 2014). Similarly, the 'sound of the MPC' gets particular attention in press and literature, a characteristic that is attributed to the lower sampling resolution of older models, resulting in lower fidelity and a dynamically limited headroom that is actually beneficial to beat 'placement' within the mix.[19] The MPC Renaissance has been chosen here as a later incarnation of this archetypical hip-hop production tool (albeit one with improved computer integration, helpful to the scope of this experiment), effectively seen as a hip-hop 'instrument' par excellence that inspires unique musical and sonic utterances.

Following experimentation with the sequence and timing of the chopped bars derived from the original composition, and making use of the MPC drum pads, it has been possible to create new rhythmical and harmonic combinations by triggering shorter segments and creating reimagined sequences that were never performed on the original recording. Depending on the length of the segments used, the tempo of the original piece was still perceptible for any sample longer than an individual percussive hit, so the whole program was detuned by two semitones, consequently reducing the tempo of the segments by approximately 12 bpm.[20] A typical boom-bap practice is to set a program's polyphony to mono, so that each segment triggered, mutes the previous one already playing. Although this was historically practiced partly as a means to obscure the origin of phonographic samples (by keeping them shorter, and presenting them in reimagined sequences), two positive side-effects of the process were a highly rhythmical effect, and preservation of clarity in the harmonic progression of newly constructed patterns (avoiding the juxtaposition of overlapping chords). Employing this practice for both aesthetic and pragmatic reasons, the following sequences were composed, stemming from the original 'chops':

TABLE 1.2 Harmonic progression of hip-hop production

Bars	*1*	*2*	*3*	*4*	*5*	*6*	*7*	*8*
Intro 1	iv iv i i	iv iv i i	iv iv i i	iv iv i i				
Intro 2	i i i iv	i i i V	i i i V	i i bVI(7) V				
Verse	iv *[x4]*	i *[x4]*	iv *[x4]*	i *[x4]*	iv(sus4) *[x2]* iv *[x2]*	i *[x4]*	iv(sus4) *[x2]* iv *[x2]*	V(7) *[x4]*
Bridge	bVI(7) *[x4]*	i *[x4]*	bVI(7) *[x4]*	i *[x4]*	bVI(7) *[x4]*	i *[x4]*	bVI(7) *[x4]*	V(7) *[x4]*
Chorus	i i iv iv	i i iv iiø	i i iv iv	i i iv iiø	i i iv iv	i i iv iiø	i i iv iv	i i iv iiø

Each chord represents one of four beats in a bar.

The large range of samples exported provided multiple alternatives for each single-chord bar, with factors such as the richness of the frequency spectrum and the individual micro-motifs of included instruments becoming crucial in the selection process and triggering. Importantly, as can be seen in Table 1.2, the programming and reorganisation of the segments allowed the construction of different sequences to the original, at altered harmonic speeds for most sections (generally faster when compared to the original recording, with speeds of two chords per bar and at times one chord per beat). It is interesting to note from the harmonic analysis that the programming and reorganisation of the sampled segments created a number of harmonic departures, extensions, or substitutions. Specifically, on the fifth and seventh bars of each verse, the segment chosen in support of the iv chord is a different one to that used in bars one and three, featuring a clear variation in the contained piano melody, and one that stresses the seventh scale degree, infusing the iv chord with a potential sus4 colouration. A similar occurrence can be observed in the chorus, where every other bar can be perceived as a half-diminished ii chord (in place of the iv chord of bars one, three, five, and seven). This is due to a passing note audible on the sampled electric bass part, which moves to the second degree of the scale, and it is the result of an additional one-beat segment brought in on beat four of every even bar (bars two, four, six, and eight of the chorus sequences). Despite the fact that these extensions and substitutions are open to interpretation, they are however *suggested* as a consequence of melodic content occurring inadvertently within the high number of sampled alternatives. The rap producer then has a choice to either exploit what is implied and augment it, or suppress it, through additional layering.[21] What's more, the sonic manipulation of sampled content within a sampler or during mixdown can further affect these harmonic choices. It is typical practice to equalise samples using a sampler's onboard filters in order to remove unwanted or clashing parts from the frequency spectrum, or to boost frequencies by picking complimentary tones that work in the new context. This can accentuate or mask particular instrumental parts literally influencing the level of their contribution to the harmonic content.

With regards to the rhythmical implications of the process, the monophonic triggering and muting, on the other hand, can create tightly syncopated results due to the placement of the new 'cut' (initiated by the percussive attack of the edit or a drum-hit on the first beat) against rhythmical subdivisions already present in the previously playing segment. In this example, this was further exploited by decimal alterations to the overall tempo, and the use of MPC's higher settings of swing quantisation, which made any sixteenth triggering 'late' and, by consequence, closer to the next event triggered. As a result, the original material here assumes new rhythmical qualities due to its placement and truncation within the programming order sequenced on the MPC. It could be argued, that the resulting sensibility is quintessentially Hip Hop: the *meta*-syncopation interacts favourably with the sampled material's internal syncopation, which may be a further—pragmatic—reason explaining why Hip Hop favours Funk and frequently cites an Afrocentric sonic past. Schloss (2014, p. 159) explains:

> A hip-hop beat consists of a number of real-time collective performances (original recordings), which are digitally sampled and arranged into a cyclic structure (the beat) by a single author (the producer). In order to appreciate the music, a listener must hear both the original interactions and how they have been organised into new relationships with each other ... And the formal structure may reflect both linear development (in the original composition) and cyclic structure (in its hip-hop utilization).

Due to legal or content-related limitations in accessing favourable sampling material, however, much rap music that is produced with cyclic priorities in mind, whilst exploiting the rhythmical tendencies described, also tends to be harmonically more timid. This is in no way a criticism of the musical outputs of the practice—to which Schloss attributes a defining aesthetic value—but it may be worth considering the creative possibilities available should this limitation be removed. The original compositional phase in this experiment has allowed for extended sampling opportunities, optimised synchronisation, and direct access to instrumental-only material. It is in the context of untapped potential that inter-stylistic evolution can be pursued further. Some of the creative answers to the ongoing debate on sampling versus live performance regarding hip-hop authenticity may lie in the grey area between these two polarities. After all, samples contain live musicianship per se, so it is the differentiating variables pertaining to the sonic domain between the sample-based method and the live approach that are of interest. Despite live hip-hop bands' best efforts to stay within the genre, the debate continues. The Roots drummer and producer ?uestlove has dedicated a large part of his professional life to achieving authentic hip-hop sonics on his drum kit, but what he may—purposely—be missing is the *meta*-ingredient: the effect of the sample-based process upon his Funk. An effect that birthed the Hip Hop of the Golden Age and defined the boom-bap subgenre. Schloss (2014, p. 151) argues:

> It is in the relationship between the samples that the process of composition begins to exert a decisive influence as producers experiment with different patterns and approaches to organization ... In making their studios into laboratories, producers are making themselves into research scientists....

Video 1.2, available from the 'Book > Videos' menu tab at www.stereo-mike.com, showcases an early study of hip-hop sections constructed out of the instrumental chopping (with embedded captions and citations).

Conclusion

This chapter has focused on the relationship between the sonic priorities of sample-based Hip Hop and the composition of original content, utilising the Blues as a case study and exploring the potential of inter-stylistic trans-morphing between its form and beat-making. Although the Blues shares some of the cyclic structures that are mirrored in Hip Hop, it has also been exploited less than other forms of music in sampling practices, thus presenting some unique rhythmical and harmonic problematics in the applied aspect of this examination. Although the chapter in no way offers an exhaustive typology of the creative opportunities that exist between the sample-based landscape and original composition, the investigation has pursued a two-fold intention: to systematically explore inter-stylistic synergies from a practice-based perspective, whilst navigating alternative creative avenues for Hip Hop's future evolution. In terms of praxis, the informed re-enactment of vintage workflows has facilitated the pre-production of a substantial body of usable source content (referential to the Blues as exemplified in this case, and beyond, as part of the larger project's undertaking), offering multiple opportunities for creative interaction. The final piece arrived at—as a result of consecutive beat-making 'studies' using the source content showcased here—is 'Left Right', further illustrating the rich potential for variation that exists through 'play' with the same raw materials.[22] Conversely, source content produced in the Blues period of the project's journey (see Figure 0.1 in the previous chapter), finds use in multiple beats comprising the end album—specifically: '1960s 2', 'It Meters', 'King's Funk', 'Negrito Blues', 'Rebluezin', and 'Take 3'. Chapter 6 will demonstrate how elements from this stylistic category of raw material can further be deployed in secondary, ornamental functions, providing supportive layers to beats that leverage content from other eras/genres as their main structures (as will be shown in the deconstruction of track 'Call'). The inter-stylistic layering in such cases becomes multi-dimensional, and the sonic implications exponential—this will be the subject of forthcoming Chapters 3–6. But first, let's delve deeper into the nuances of the sampling technology that makes all this possible.

Video 1.3, available from the 'Book > Videos' menu tab at www.stereo-mike.com, illustrates the 'Left Right' beat with recorded rap layers.

Recommended chapter playlist
(in order of appearance in the text)

'Left Right'
'1960s 2'
'It Meters'
'King's Funk'
'Negrito Blues'
'Take 3'
'Rebluezin'
'Call'

Notes

1 A *break* or *break-beat* refers to the rhythmical breakdown of a record occupied solely by drums. DJs would extend the break-beat's duration using two copies of the record on two turntables and switching continuously between the two breakdown segments. MC stands for Master of Ceremony, and later Microphone Controller, both referring to rapper in this context.

2 Another heavily sampled drum break is the intro from Led Zeppelin's 'When The Levee Breaks' from *Led Zeppelin IV* (1971)–ironically a Memphis Minnie cover, bringing us full circle back to the Blues.

3 Interpolation refers to the studio re-creation of performances and sonics of an existing recording, which avoids breaching mechanical (phonographic) copyright, whilst still in use of the original composition (publishing rights).
4 This is a delineation Williams (2010, p. 21) describes—citing Lacasse—as "autosonic" versus "allosonic" quotation, respectively.
5 Four studios have been used for recording (DARP Studios in Atlanta; Electric Lady Studios; and Sony Music Studios in New York) and mixing the track (Circle House Studios in Miami).
6 In an interview with the author, Gazaway (2016) details how he accesses the needed samples for his "conceptual collaborations" from multiple sources: available multitracks, isolating the left and right sides of a stereo master, locating extended live performance versions, and sampling solos and exposed instruments from these; furthermore, he purposely samples B.B. King's *Lucille* guitar and treats it as a separate character in his arrangements.
7 The album that is credited with putting Southern Rap on the map is OutKast's *Southernplayalisticadillacmuzik* (1994), complete with live performances of slowed-down southern soul meeting synthetic drum-machine programming (see, for example: Grem, 2006).
8 Gazaway (2016) regards DJ Screw and Pimp C as pioneers of the Southern Rap sensibility, a notion that is echoed by the hip-hop community at large.
9 Beer and Sandywell (2005, p. 115) define trans-morphing as "the creation of trans-genres by morphing across genres ... This process generates a hybrid genre as the performer is simultaneously positioned in two or more genres".
10 The track further abuses the pitch-tempo relationship at 5:12, down to Fm and 65 bpm.
11 'Spill' in studio recording refers to the leakage of sonic content reaching a microphone positioned closest to the intended source from surrounding instruments.
12 DJs momentarily mute records for rhythmical effect using a crossfade control on their mixer during live performances, and in the practice of turntablism this is referred to as a 'cut'; it is often emulated on studio recordings by automation, using the mute button on a mixing console, or via various editing practices in software.
13 Evans (2000, p. 90) discusses various manifestations of this very characteristic when comparing Blind Lemon Jefferson with his contemporaries; highlighting Jefferson's innovations he details:

> Jefferson's practice of prolonging the singing of certain notes and thereby stretching the standard twelve-bar form is illustrated in virtually all of his blues using an AAB stanza pattern. In these he also contributes to the stretching by playing extended guitar figures in response to his vocal lines.

14 Although recording to a click-track is atypical of blues sessions of the referenced era, the rationale behind this decision has been to aid sampling on a much larger scale than typically practiced. For most sample-based rap productions, a smaller number of samples are chosen from the same record and the variations in timing on the original performances can be negotiated though time-based manipulation. But for the 125 samples chosen from the original here, some degree of synchronisation had to be maintained for the sampling phase to remain feasible.
15 The references range from the predominantly straight-eighth patterns audible on the Freddie King and B.B. King examples discussed, to the more swung-sixteenth patterns performed by Al Jackson Jr. for Albert King at Stax.
16 The recording process was designed to honour tracking practices and appropriate instrumental sources representative of the era. The resulting channel list is as follows:
 1. Bass drum
 2. Snare drum
 3. Middle tom
 4. Mono overhead (option)
 5. Stereo overhead (left)
 6. Stereo overhead (right)
 7. Drum room microphone
 8. Shaker
 9. Fretless bass
 10. Rhodes
 11. Upright piano (left)
 12. Upright piano (right)

13. Clean electric guitar (rhythm)
14. Fuzz electric guitar (rhythm)
15. Clean and fuzz electric guitar (lead)

17 The MPC Renaissance is a descendant of 1988s Akai MPC 60 and a mainstay in the current arsenal of hip-hop production tools.

18 The MPC operating script uses programs as groupings of multiple samples, sharing a number of user-definable parameters, such as polyphony, effects, and output assignments (see Chapter 2 for more on the relationship between the technical specifications of the MPC range and aesthetic/stylistic implications for Hip Hop).

19 The company itself pays tribute to its heritage (and sampling heritage in general) by including four options for modelling vintage sampler behaviour: that of the Akai MPC 3000, the MPC 60, and two variations for the E-mu SP-1200.

20 This echoes Gazaway's method described above, but it is also typical of hip-hop practice in general, as a means to further distance one's output from the recognisability of the source and arrive at tempi appropriate for the subgenre in question: here, the vision of a more harmonically rich hip-hop production lends itself to a Southern Rap sensibility, which frequents slower tempi.

21 In the case on hand, a Hammond organ part has been added in post-production to support these harmonic 'suggestions'.

22 A practice that exemplifies Schloss's (2014, p. 159) earlier point of "original interactions [*perpetually*] organised into new relationships".

Bibliodiscography

Abdominal and The Obliques (2012) *Sitting Music* [CD, Album]. Canada: Pinwheel Music.

Beastie Boys (2022) *Like making a two year old pizza, Instagram*. Available at: https://www.instagram.com/p/ChDbTJ6jB8H/ (Accessed: 12 March 2023).

Beer, D. and Sandywell, B. (2005) 'Stylistic morphing: Notes on the digitisation of contemporary music culture', *Convergence: The International Journal of Research into New Media Technologies*, 11(4), pp. 106–121.

Bernstein, A. (2016) 'Interviewed by author, 21 May'.

Blakroc (2009) *Blakroc* [Vinyl LP]. UK & Europe: BlakRoc Records.

Brown, J. (1970) *Funky Drummer* [Vinyl, 7"]. US: King Records.

Caldwell, B. (2015) *The Awesome B.B. King/UGK Mashup, Houston Press*. Available at: https://www.houstonpress.com/music/ugk-bb-king-mashup-the-trill-is-gone-is-as-awesome-as-you-think-it-is-7887211 (Accessed: 1 December 2020).

Chang, J. (2007) *Can't stop won't stop: A history of the hip-hop generation*. Reading, PA: St. Martin's Press.

Chang, V. (2009) 'Records that play: The present past in sampling practice', *Popular Music*, 28(2), pp. 143–159.

Chuck D (2003) 'Blues: The footprints of popular music', in P. Guralnick, R. Santelli, and H. George-Warren (eds) *Martin Scorsese presents the Blues: A musical journey*. New York: Harper Collins, pp. 280–281.

Collins, S. (2008) 'Waveform pirates: Sampling, piracy and musical creativity', *Journal on the Art of Record Production*, 3.

Evans, D. (2000) 'Musical innovation in the Blues of Blind Lemon Jefferson', *Black Music Research Journal*, 20(1), pp. 83–116.

Gazaway, A. (2015) *B.B. & The Underground Kingz: The Trill Is Gone* [Digital Release, Album]. Soul Mates Records.

Gazaway, A. (2016) 'Interviewed by author, 11 August'.

George, N. (2005) *Hip hop America*. London: Penguin Books.

Goodwin, A. (1988) 'Sample and hold: pop music in the digital age of reproduction', *Critical Quarterly*, 30(3), pp. 34–49.

Grem, D.E. (2006) '"The South got something to say": Atlanta's dirty South and the southernization of hip-hop America', *Southern Cultures*, 12(4), pp. 55–73.

Guralnick, P., Santelli, R. and George-Warren, H. (2004) *Martin Scorsese presents the Blues: A musical journey*. New York: Harper Collins.

Harkins, P. (2008) 'Transmission loss and found: The sampler as compositional tool', *Journal on the Art of Record Production*, 4.

Harkins, P. (2010) 'Appropriation, additive approaches and accidents: The sampler as compositional tool and recording dislocation', *Journal of the International Association for the Study of Popular Music*, 1(2), pp. 1–19.

Harling, D. (2013) *Bun B Breaks Down Origins of 'Trill' With A$AP Rocky, HipHopDX*. Available at: https://hiphopdx.com/news/id.25756/title.bun-b-breaks-down-origins-of-trill-with-aap-rocky (Accessed: 1 December 2020).

Hawkins, R. (2000) The Thrill Is Gone: The Legendary Modern Recordings [CD, Album]. UK: Ace.

King, A. (1972) *I'll Play The Blues For You* [Vinyl LP]. US: Stax.

King, B.B. (1969) *The Thrill Is Gone* [Vinyl LP]. US: Bluesway.

King, F. (1973) *Woman Across The River / Help Me Through The Day* [Vinyl, 7"]. US: Shelter Records.

Kulkarni, N. (2015) *The periodic table of Hip Hop*. London: Random House.

Led Zeppelin (1971) *Led Zeppelin IV* [Vinyl LP]. Europe: Atlantic.

Levin, M. (2004) 'Godfathers and sons', in P. Guralnick, R. Santelli, and H. George-Warren (eds) *Martin Scorsese presents the Blues: A musical journey*. New York: Harper Collins, pp. 186–187.

Marshall, W. (2006) 'Giving up hip-hop's firstborn: A quest for the real after the death of sampling', *Callaloo*, 29(3), pp. 868–892.

McLeod, K. (1999) 'Authenticity within hip-hop and other cultures threatened with assimilation', *Journal of Communication*, 49(4), pp. 134–150.

Nas (2004) *Street's Disciple* [CD, Album]. Europe: Sony Urban Music.

OutKast (1994) *Southernplayalisticadillacmuzik* [CD, Album]. Europe: LaFace Records.

Quinlan, T. (2012) *Abdominal & the Obliques: Sitting Music, Exclaim*. Available at: https://exclaim.ca/music/article/abdominal_obliques-sitting_music (Accessed: 1 December 2020).

Reiff, C. (2015) *Texas rap meets Memphis blues on the B.B. King/UGK mashup LP the Trill Is Gone, AV Club*. Available at: https://news.avclub.com/texas-rap-meets-memphis-blues-on-the-b-b-king-ugk-mash-1798285573 (Accessed: 1 December 2020).

Roberts, R. (2012) *Weekend mixtape: The Pharcyde collides with Tribe Called Quest, Los Angeles Times*. Available at: https://www.latimes.com/entertainment/music/la-xpm-2012-sep-14-la-et-ms-weekend-mixtape-pharcyde-tribe-called-quest-20120914-story.html (Accessed: 23 August 2021).

Rodgers, T. (2003) 'On the process and aesthetics of sampling in electronic music production', *Organised Sound*, 8(3), pp. 313–320.

Rose, T. (1994) *Black noise: Rap music and black culture in contemporary America*. Hanover, NH: University Press of New England (Music/Culture).

Rothenbuhler, E.W. (2006) 'For-the-record aesthetics and Robert Johnson's blues style as a product of recorded culture', *Popular Music*, 26(1), pp. 65–81.

Schloss, J.G. (2014) *Making beats: The art of sample-based Hip-Hop*. Middletown, CT: Wesleyan University Press (Music/Culture).

Serrano, S., Torres, A. and Ice-T (2015) *The rap year book: The most important rap song from every year since 1979, discussed, debated, and deconstructed*. New York: Abrams Image.

Soul Mates Records (2015) *B.B. & The Underground Kingz: The Trill Is Gone, by Amerigo Gazaway, Bandcamp*. Available at: https://soulmatesproject.bandcamp.com/album/b-b-the-underground-kingz-the-trill-is-gone (Accessed: 23 August 2021).

Swiboda, M. (2014) 'When beats meet critique: Documenting hip-hop sampling as critical practice', *Critical Studies in Improvisation*, 10(1), pp. 1–11.

Toop, D. (2000) *Rap attack 3: African Rap to global Hip Hop*. 3rd edn. London: Serpent's Tail.

Various (1996) *The Roots of Rap: Classic Recordings From The 1920's And 30's* [CD, Album]. US: YAZOO.

Waters, M. (1955) *Manish Boy* [Vinyl, 7"]. US: Chess.
Waters, M. (1968) *Electric Mud* [Vinyl LP]. UK: Chess Records.
Williams, J.A. (2010) *Musical borrowing in hip-hop music: Theoretical frameworks and case studies*. Unpublished PhD thesis. University of Nottingham.
Williams, J.A. (2014) *Rhymin' and stealin': Musical borrowing in Hip-Hop*. Ann Arbor: The University of Michigan Press.
Zak III, A.J. (2001) *The poetics of rock: Cutting tracks, making records*. Berkeley: University of California Press.

2

BOOM-BAP AESTHETICS AND THE MACHINE

Over the past few decades, the growing musicological literature on Hip Hop has paid ample tribute to Akai's range of MPCs (originally, MIDI production centres—currently, music production controllers), acknowledging their pivotal influence on rap production practices (Rose, 1994; Rodgers, 2003; George, 2005; Harkins, 2008, 2010; Williams, 2010; Sewell, 2013; Morey and McIntrye, 2014; Ratcliffe, 2014; Schloss, 2014; Swiboda, 2014; Wang, 2014; D'Errico, 2015; Kajikawa, 2015b; McIntyre, 2015; Shelvock, 2017). The technology's combined functionality of sampling, drum machine, and MIDI-sequencing features has been embraced by rap practitioners ever since the release of the standalone MPC 60 in 1988. The timeline coincides with particular sonic priorities in Hip Hop that can be grouped under the *boom-bap* aesthetic—an onomatopoeic celebration of the prominence of sampled drum sounds programmed over sparse and heavily syncopated instrumentation. But what is the association between subgenre aesthetics and MPC functionality, and what parallels can be drawn between the evolution of the technology and stylistic deviations in the genre? This chapter examines how MPC technology impacts upon the stylisation of Hip Hop as a result of unique sonic, rhythmic, and interface-related characteristics, which condition sampling, programming, and mixing practices, determining in turn recognisable sonic signatures. Furthermore, the boom-bap sound is traced from its origins in the mid- to late-1980s, through to its current use as an East Coast production reference, honouring a sample-based philosophy that is facilitated by the MPCs' physical interface and operating script. The findings from a number of representative case studies form a systematic typology of technical characteristics correlated to creative approaches and resulting production traits, informing speculation about the future of the MPC, its technological descendants, and the footprint of its aesthetic on emerging styles and technologies. What follows links the creative opportunities uncovered in the previous chapter, to the material priorities afforded by the technology's specific design features.

DOI: 10.4324/9781003027430-5

A quick rewind <<

It was 25 years ago when *Keyboard* magazine made sampling front-page news, featuring DJ Shadow on the front cover alongside the Akai MPC 2000, and leading with the story 'Samplers Rule'. Jim Aikin (1997, p. 47) wrote:

> When a new technology gets loose in the music world, it takes a few years to become front-page news. Consider the electric guitar... The sampler is making a serious bid to be the electric guitar of the '90s... What has reached critical mass is the complex of music (and social) meanings attached to sampling.

The story of sampling, however, had begun over a decade and a half before (at this point, the MPC 2000 represented the fourth generation of MPC technology), and Hip Hop's origins can be traced prior to the invention of digital sampling technologies altogether. Rap aficionados associate hip-hop music automatically with sampling, but to put things in perspective, Hip Hop's original 'instrument' was the turntable (Katz, 2012, pp. 43–69). As seen in Chapter 1, pioneering DJs in the Bronx had used a pair of turntables and a mixer to extend the instrumental sections of the 1960s and 1970s soul and funk recordings, and this is what provided the sonic foundation for MCs to eventually rap over (Toop, 2000; Chang, 2007). But it took a number of years before any rapping was actually committed to record and the first successful rap release arrived in the form of *Rapper's Delight* (1979) by the Sugarhill Gang (Howard, 2004). *Rapper's Delight* and all proto-rap releases of the era utilised disco, soul, and funk musicians for the production of the instrumental backing (Kulkarni, 2015; Serrano, Torres and Ice-T, 2015), and turntables did not feature prominently on phonographic records until *The Adventures of Grandmaster Flash on the Wheels of Steel* (1981). Neil Kulkarni (2015, p. 37) informs us that: "In late 1982 and early 1983, hip-hop records didn't sound like hip-hop. They were essentially R'n'B records with rapping on them, created by bands, session players and producers". Flash's turntable performance stands both as a historical *record* of the performative hip-hop tradition (turntablism), as well as a music production that contains multiple phonographic segments that are then cut, manipulated, and juxtaposed further by the DJ. It is these sonic artefacts that early hip-hop producers attempted to replicate when the first affordable digital samplers hit the market—a notion that is particularly audible in mid- to late-1980s sample-based releases. Ingenuity on the side of the beat-makers and evolving design on the side of the manufacturers meant that samplers would soon transcend the function of merely replicating turntable performance, unpacking new creative possibilities, and becoming hip-hop instruments par excellence (in the hands of studio DJs who were transitioning to fully fledged *producers*).

The boom-bap signature

The story of Boom Bap is closely associated with the development and practice of sampling, and as such, Boom Bap is often described as a production technique, a sound, a style, or a subgenre. The term was first uttered by T La Rock in the final ad-libs of 'It's Yours' (1984) (Mlynar, 2013), but it was popularised by KRS-One with the release of the *Return of the Boom Bap* (1993) album. It stands for an onomatopoeic celebration for the sound of a loud kick drum ('boom') and hard-hitting snare ('bap') exposed over typically sparse, sample-based instrumental production. Although Rose (1994, p. 80) (mis)interprets a range of sampling and drum-machine technologies as "samplers", the following quotation from *Black noise* nevertheless powerfully communicates the rationale behind the prominence of the 'boom':

> Rap's engineering and mixing strategies address ways to manage and prioritize high-volume and low-frequency sounds. Selected samplers carry preferred "sonic booms" and aid rap producers in setting multiple rhythmic forces in motion and in recontextualizing and highlighting break beats. These strategies for achieving desired sounds are not random stylistic effects, they are manifestations of approaches to time, motion, and repetition found in many New World black cultural expressions.

It could be argued that the words *boom* and *bap* conjure rhythmic, timbral, and balance implications, and as such, Boom Bap may be better described as an overarching aesthetic that signifies hip-hop eras, production preferences, sonic traits, subgenre variations, geographical connotations, and even authenticity claims. Mike D'Errico (2015, p. 281) defines "boombap" as a "sound that was shaped by the interactions between emerging sampling technologies and traditional turntable practice" by producers who "used turntables alongside popular samplers such as the Akai MPC and E-Mu SP-1200", resulting in "gritty, lo-fi audio qualities… and innovative performance practices that continue to define the sound of 'underground', 'old-school' hip-hop".

But how do we make the transition from particular mechanistic affordances (Clarke, 2005, pp. 36–38) to a complex set of sonic signatures claiming their very own raison d'être?[1] The connection lies in the sampling affordances that enabled the separation, reinforcement, and stylisation of individual drum sounds within a hip-hop context to such an extent that practitioners 'baptised' the phenomenon with its own onomatopoeia. The significance of this is that the sonic variables that characterise Boom Bap are interrelated to production techniques and workflow approaches conditioned by technical characteristics found in digital samplers in general and the MPC range in particular.

Boom Bap ex machina

The isolation of the 'boom' and the 'bap' can be traced back to pioneering hip-hop producer Marley Marl, who "discovered the power of sampling drums by accident during a Captain Rock session" (Weingarten, 2010, p. 22) and, in his own words, found that he "could take any drum sound from any old record, put it in [t]here and get that old drummer sound" (cited in George, 1998/2005, p. 92). Kajikawa (2015b, pp. 164–165) informs us that Marl must have "first experimented with sampled drum breaks in or around 1984 when the first devices with adequate memory and function, such as [the] E-mu Emulator II and the Ensonique Mirage, began hitting the market". The significance of this discovery—and Marl's influence on a genealogy of producers associated with Boom Bap, such as DJ Premier, Pete Rock, Q-Tip, RZA, Prince Paul, DJ Shadow, J Dilla, and Madlib—is that it empowered beat-makers to transition from 'surface manipulators' (users of drum loops or breaks referential to a turntable affordance) to drum 'scientists': sampling-programmers who could come up with new patterns altogether, layer multiple drum sounds upon one another, and create original rhythms out of minimal sonic segments from the phonographic past. Soon, the techniques advanced to a dense layering of sampled *and* synthesised sources (the latter often courtesy of a Roland TR-808 drum machine), complex rhythmic appropriation, intricate chopping, and sound-object juxtaposition.

It is in the trajectory of this evolving production technique—from Marl's drum-hit isolation to later producers' intricate juxtaposition—that the development of the boom-bap aesthetic can be observed, highlighting the tension between 'liveness' and rigidity; organic and synthetic sonics. Talking about the Roland TR-808 and E-mu SP-1200, Kulkarni (2015, p. 43) observes that: "The two most emblematic pieces of hardware hip hop has ever used both, in their way, crystallise that delicious dilemma, that tightrope between looseness/'feel' and machine-like tightness that hip hop's sound so engagingly steps on". Naturally, with powerful sampling technology integrated alongside drum-machine and sequencer functionality within standalone music production centres, future producers would go on to approach all past phonographic material—not just funk and soul drum breaks—with increasing microscopic focus, separating instrumental phrases into short sonic 'stabs', assigning them to their drum pads, and performing and programming reimagined sequences into new cyclical arrangements (loop-based compositions). The 'boom' and the 'bap' would evolve to represent not only a drum-inspired onomatopoeia, but an overarching chopped, manipulated, and syncopated *aesthetic* founded upon the interaction of past records with new mechanistic sequencing. Table 2.1 maps distinct characteristics of the boom-bap sound against affordances—and limitations—found specifically on the MPC range. The left column highlights characteristic stylisations that define the boom-bap aesthetic, while the right column indicates software and hardware functionality within the MPC environment that promotes these stylisations or makes them possible. It is worth noting that many of these affordances are not exclusive to MPC technology anymore, but their combined integration on a standalone piece of hardware as early as 1988 became instrumental in allowing the aesthetic to develop, whilst also conditioning future workflow preferences mirrored in later generations of the same hardware (and competitive designs too). As such, the MPC workflow empowered 1990s hip-hop producers to perfect the sample-based artform, and the boom-bap sound became synonymous with Hip Hop's Golden Age (circa 1988–1998), as well as the East Coast's rather dogmatic reliance on sampling phonographic sources.

East Coast producers continued to rely on boom-bap methods not only due to their preference for phonographic raw material, but also as a reaction to the more synthesised subgenres coming out of the West Coast or the US South—a form of conscious sonic signposting towards the birthplace of the genre, New York. The boom-bap sound would later be taken into more experimental, instrumental frontiers (often referred to as 'Glitch—or Lo-Fi—Hop') by producers such as J Dilla, Madlib, Prefuse 73, and Flying Lotus (Hodgson, 2011), while in the last decade it has been enjoying a resurgence in the form of a plethora of underground and crossover releases classified as Boom Bap;[2] as well as mainstream releases increasingly tapping into it to support more conscious lyrical themes (for example, Jay-Z's 'The Story of O.J.' (from *4:44*, 2017), produced by No I.D.). In an interview with *Rolling Stone* magazine, No I.D. (cited in Leight, 2017) sums up the rationale behind his return to a sample-based approach by admitting:

> I began to play the samples like I would play an instrument … I had stepped away from my strength sometimes because the business makes you think you can't do it … I can do it. And I can create new art.

TABLE 2.1 A mapping of boom-bap stylisations against MPC affordances and limitations

	Boom-bap characteristics	*MPC affordances / limitations*
Balance	Prominent kick drum	Internal mix functionality
	Prominent snare drum	Internal mix functionality
Timbre	Emphasised low-end (kick drum)	Internal processing (effects) / Resolution limitation
	Hard-hitting snare drum (presence)	Internal processing (effects) / Resolution limitation
	Low fidelity	Resolution limitation
	Vinyl (sample) sources	Phono inputs (and preamps)
Dynamic	Compressed instrumental production	Internal processing (compression) / Resolution limitation
	Interpolation/filtering	Controllers (interface)
Sonic 'Glue'	Instrumental production 'glue'	Resolution limitation / Converters (I/O) / Internal processing (compression)
	Shared ambience on sampled elements	Internal processing (effects)
Arrangement	Isolated drum 'hits'	(Auto-)Slice functionality / Memory limitation
	(Short) Other instrumental 'stabs'	(Auto-)Slice functionality / Memory limitation
	Layered kick drum (often with 808)	Program functionality / MIDI out
	Layered snare drum	Program functionality / MIDI out
	'Chopped' breaks (drums) and other phonographic samples	(Auto-)Slice functionality
	Sparse instrumentation	Memory limitation / Program functionality (mono)
	Turntable effects/performance	Phono inputs
	Four-measure repetition / chorus variation	Sequencer / song functionality
Rhythmic	(Highly) Swung programming	MPC swing/quantisation algorithm
	Tight drum-instrumental syncopation	Program functionality / MIDI out / MPC swing/quantisation algorithm
Motivic	Re-arranged phrases/rhythms/motifs	(Auto-)Slice functionality / Drum pads (interface)
	Percussive programming of instrumental phrases	Drum pads (interface) / Program functionality (mono)

Weapon of chop

It is important, however, to consider the point at which MPCs enter the historical timeline and the rationale behind them replacing E-mus as preferred weapons of—Hip Hop—choice. Akai released the MPC 60 in 1988, bringing a number of improvements to the notion of integrated sampling, drum machine, and MIDI-programming functionality. Ratcliffe (2014, p. 113) explains:

> MPC is the model designation for a range of a sampling drum machine/sequencers, originally designed by Roger Linn and released by Akai from 1988 onwards (for

> instance, the MPC 60, MPC 2000, and MPC 3000). These instruments are favoured for sample-based hip-hop and EDM due to both the design of the user interface (featuring drum pads for real-time programming) and idiomatic performance characteristics (such as the swing quantisation algorithm).

Rap producers made the switch from E-mus to Akais for different reasons, but it could be summarised that the MPCs' unique swing quantisation parameter, higher bit-depth resolution, the touch-sensitive drum pads of their physical interface, and the internal mixing functionality were among the main reasons (Anderton, 1987; Linn, 1989). Roger Linn (Scarth and Linn, 2013) himself attributes the "natural, human-feeling grooves in [his] drum machines … (i)n order of importance" to the factors of: "(s)wing"; the "(n)atural dynamic response on [the] drum pads"; the "(p)ressure-sensitive note repeat" function; programming accuracy; strong factory sounds; and a user-friendly interface. Hank Shocklee (cited in Dery, 1990, pp. 82–83, 96) concurs:

> [The 1200] allows you to do everything with a sample. You can cut it off, you can truncate it really tight, you can run a loop in it, you can cut off certain drum pads. The limitation is that it sounds white, because it's rigid. The Akai Linn [MPC-60] allows you to create more of a feel; that's what Teddy Riley uses to get his swing beats.

Schloss (2014, pp. 201–202) adds that:

> the circuitry and programming of different models of samplers are believed to impart special characteristics to the music (perhaps the best known of these characteristics is the legendary 'MPC swing', a rhythmic idiosyncrasy first noted in the Akai MPC 60 sampler, circa 1988).

And Kajikawa (2015a, p. 305) reports that the Akai MPC 60's "touch-sensitive trigger pads allowed producers to approach beat-making with renewed tactile sensitivity". Figure 2.1 provides a schematic representation of a timeline, mapping stylistic hip-hop eras against the manufacture of particular models of E-mu and Akai products, alongside examples of seminal phonographic hip-hop releases.

A typology

The technical characteristics of the MPC range can thus be grouped into variables relating to the operating script on the one hand, and to the physical attributes of the hardware interface on the other. These, in turn, influence sampling, programming, and mixing tendencies in producers' workflows, with sonic, rhythmic, and motivic implications for the musical outputs. It is beyond the scope of this chapter to detail every function, physical attribute, or parameter found on the MPC range, so the focus will remain instead on characteristics that create noteworthy affordances in producers' workflows. These are, in turn, mapped to a number of predictable sonic signatures (potential aesthetic results), as can be observed through aural analysis of seminal works. Consequently, the observations here do not follow a one-way, technologically deterministic purview, but take into account the creative agency of producers illuminated through the discussion of key works, and informed by existing

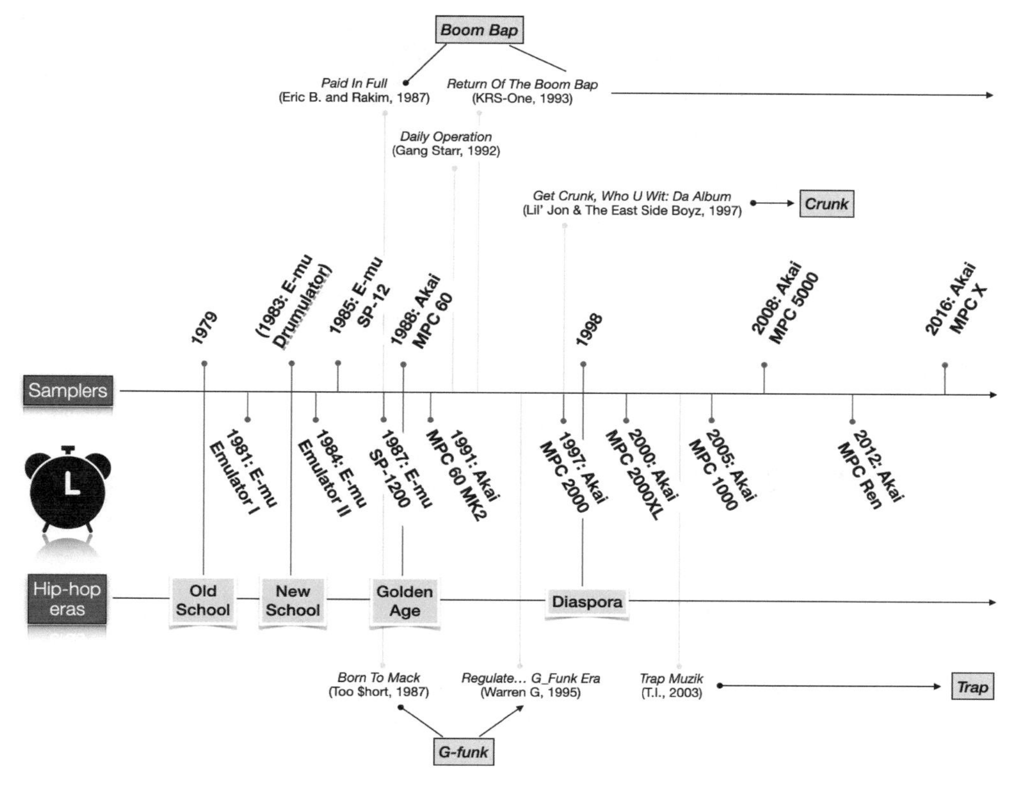

FIGURE 2.1 Timeline mapping hip-hop eras against E-mu and Akai products, with examples of seminal releases characteristic of rap subgenres.

musicological literature and producer testimonials. A visual representation of the typology is provided in Figure 2.2.

Starting from the MPC's operating script, key characteristics highlighted in the typology are: (a) the quantisation algorithm (and MPC's infamous 'swing' parameter); (b) the MPC's onboard sequencer, its looping function, and the included 'song-mode' for the construction of longer phrases; (c) the note-repeat function (tied to the sequencer and quantisation function); (d) the (auto-)slice option of the zone functionality (introduced in 1997 with the release of the MPC 2000), which enables the separation of a longer audio sample into separate segments or 'chops'; (e) the memory limitations of earlier designs (resulting in shorter sampling times); (f) the monophonic/muting functionality within programs (as deployed for the blues audio examples of the previous chapter); and (g) the internal routing functionality (including optional effects boards introduced after 1997 with the release of the MPC 2000).

Key features of the physical interface are: (a) the velocity-sensitive drum-style finger pads; (b) physical controllers such as sliders and rotary knobs found on the hardware; and (c) various aspects of the MPCs' input/output (I/O) functionality, which can be further subdivided into: (i) the type of sampling inputs available; (ii) the (multiple) outputs functionality (and how this correlates to internal routing and processing); and (iii) the quality and resolution

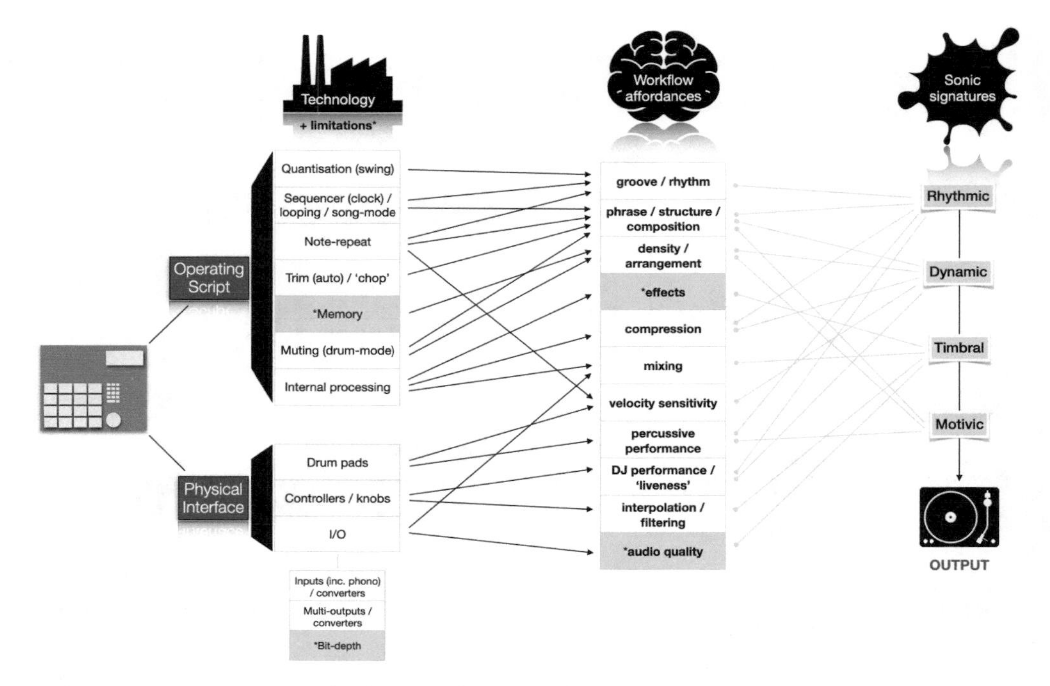

FIGURE 2.2 A schematic representation of technical characteristics of the MPC range mapped against workflow affordances and sonic signature categorisations.

of the analogue-to-digital (A/D) and digital-to-analogue (D/A) conversion, including the bit-depth limitations of earlier designs (and their emulation thereof in later ones).

Next, I turn my attention to how these technical features have affected creative (ab)use in the context of hip-hop music making and what their influence has been on the boom-bap sound. To aid the investigation, it is worth revisiting some seminal work in the genre and examining the sonic signatures present in relation to the highlighted features. For a comparative analysis, I consider two releases before and after 1988 that have been classified as Boom Bap: Eric B. and Rakim's album *Paid in Full* (1987), largely produced by Eric B. but influenced by Marley Marl, and featuring two of his remixes (partly recorded at his home studio); and Gang Starr's *Daily Operation* (1992), featuring DJ Premier's archetypal—and highly swung—production footprint.

Rhythmic signatures

Comparing Marl's remixing on 'My Melody' or 'Eric B. Is President', to DJ Premier's programming on tracks such as '2 Deep' or 'Take It Personal', there is a discernible difference evident in the swing quantisation of various elements. We know for a fact that although DJ Premier learned his craft on an E-mu SP-12, by the early 1990s he had switched to an Akai MPC 60 for his programming workflow, triggering samples from an Akai S 950 sampler (Tingen, 2007). Based on previous research (George, 2005; Weingarten, 2010; Kajikawa, 2015b), it is also safe to assume that what we are hearing on *Paid in Full* is E-mu technology, especially as this is the year prior to the release of the MPC 60. Marl's work is indeed swung (as can be clearly heard in the rhythmic placement of the sixteenth kick-drum figure in the third measure of each four-measure loop on 'Eric B. Is President'), but when compared to DJ Premier's programming (on elements such as the brass stabs of '2 Deep' and the swung kick-drum sixteenths on 'Take It Personal'), the latter is so highly swung it almost resembles a triplet feel. In fact, it is impossible to find any records prior to 1988 that are as highly swung as DJ Premier's work in the early 1990s, a fact that ties this characteristic musical figure to the MPC swing algorithm. But the sonic artefacts observed here cannot be attributed to the technology alone. DJ Premier's love of jazz (as exemplified by his sampling choices famously featuring Charlie Parker on Gang Starr's debut release *No More Mr Nice Guy* (1989)) surely has a lot to do with his abuse of the swing affordance, and it is precisely this flux between absorbed influences (culture), technology, and personal agency that results in stylistic evolutionary leaps.

Such was the effect of MPC quantisation parameters on hip-hop outputs that in the following years producers would meticulously reap the time intricacies of different generations of the hardware and import them into their digital audio workstations, emulating the MPC 'feel' in absentia. Long before these rhythmic signatures came pre-packaged in contemporary DAWs such as Ableton Live, the beat-making community would share them online (in the form of MIDI files and song templates). Certain producers would even extract them in person to ensure higher rhythmic accuracy between sequenced elements on their version of the hardware and programmed elements running in parallel on their DAWs. Celebrated producer Just Blaze (cited in Gallant and Fox, 2016) is one of the beat-makers who made the switch from producing on hardware MPCs to Apple's Logic DAW, and in an interview with uaudio.com he provides valuable insight into his rationale:

> When I made the decision to move over to Logic, a couple of guys that I work with and I imported all of the actual sequencer grooves from three MPCs into Logic ...
>
> We did this because, even though it's all ones and zeros and they're all computers, every processor and sequencer is a bit different... The grooves that we imported over are the only thing that I always make sure that I have warmed up when I open a blank session.

Just Blaze here highlights another important quality of the MPC range (and hardware/sampling drum machines in general): the computerised 'grid', stipulated by the selected tempo and its subdivisions, is not actually set in stone, and timing idiosyncrasies in hardware sequencers do not stop with swung quantisation templates. It has been shown that even when a straight feel is selected, an MPC imparts timing errors that deviate from a strict mathematical grid, and the complexity of the programmed material can have an incremental effect on timing accuracy (Perron, 1994, p. 5).

Dynamic signatures

The combination of unique quantisation templates and timing idiosyncrasies can therefore be seen as a useful binary of control versus randomisation, which, in the right programming hands, allows for the creation of 'groove' and 'feel' signatures, balancing the quest for rhythmic tightness with Hip Hop's 'genetic' predisposition for a 'live' heartbeat. But no rhythmic analysis of groove is complete without a mention of dynamics, and as any drummer (or drum programmer) would likely concur, rhythmic feel is the result of not just timing but also accents and velocity variations. Here, the touch-sensitive drum pads of the physical interface enable further interaction between human and machine-like qualities. The drum pads on the MPC can be used in a fully touch-sensitive mode, registering any velocity that the performer/programmer has exerted as a MIDI value; or, alternatively, they can be pre-set to threshold or stepped velocities, allowing for more controlled expression. Furthermore, the note-repeat function on MPCs enables the automatic repetition of a sample assigned to a drum pad according to the pre-set quantisation value, but the benefit of touch-sensitivity again presents a unique opportunity for mechanistic timing over expressive dynamics.

Motivic signatures

An important affordance that enabled further agency with the drum break—and closer interaction with individual 'stabs' or 'hits'—was the auto-slice option implemented as part of the zone functionality since the release of the MPC 2000 (Avgousti, 2009). Beat mavericks like J Dilla and Madlib clearly abused this function on albums such as *Champion Sound* (Jaylib, 2003), but—in fact—this automated process mirrored a practice long-exercised by boom-bap producers. Whether laboriously chopping longer drum breaks (or motivic phrases) 'by hand' or using automated processes to separate them into shorter segments, beat-makers took pride in meticulously subdividing sections to the smallest temporal denominator necessary—eighths, sixteenths, or even individual hits or stabs—before assigning these 'cuts' onto their drum pads for re-triggering. Although this process may now feel quite commonplace to the contemporary producer, the rhythmic, dynamic, and motivic implications of this practice on the stylisation of hip-hop production at the time were of massive importance. It allowed for increased rhythmic freedom, intricate recontextualisation of source material, and striking syncopation, but it also had timbral implications as will be shown next, especially when used in conjunction with the MPCs' program-muting/monophonic mode.

In Chapter 1, we saw how organising audio segments from a blues production, as samples belonging to the same program within the MPC environment, allowed for a number of characteristic boom-bap stylisations. It is worth delving a little bit deeper into the MPC's operating script to clarify how these come to fruition. Firstly, it is helpful to understand the hierarchical organisation of assets as part of a typical MPC beat-making workflow. Beat-makers tend to organise their MPC assets (which can include recorded samples, programmed sequences, effect configurations, virtual mixer setups, and output assignments) under projects or songs. Projects or songs can include numerous sequences, programmed or performed by the beat-maker using the MPC's drum pads and/or controllers (and progressively in newer models also by entering and manipulating onscreen graphical representations of note/controller events within sequences). Sequences, in turn, contain multiple tracks, which adhere to the sequence's chosen tempo and time signature. And each track can be assigned a program that involves a unique configuration of recorded samples, including parameters that affect their performance modes and sonic character (audio effects and output/routing configuration). Some of these variables can be dynamically controlled by assigning samples to a program's zones—an option particularly favoured when working with older MPC models (with limited memory) to allow for parameter variation economically deploying a smaller number of programs (i.e. avoiding unnecessary duplication of programs merely due to the need for parameter differentiation). The tracks, with their assigned programs, resemble elements of other multitrack media (albeit in a virtual production environment), and therefore a track-based mixer configuration within the MPC is needed to exercise control over their relative balance, levels, inserted effects, auxiliary sends to parallel effects, output grouping and processing, and automation of mixer parameters. Summarising the hierarchy, it can be typically considered as: project/song > sequence/mixer > track > program > (zone >) sample.

Key to the percussive/syncopated mobilisation of source objects described as *quintessentially Hip Hop* in the previous chapter is the combination of two parameters: setting samples to 'one-shot' mode, which ensures that they play throughout the entire duration of an established/truncated segment (i.e. a 'chop') once triggered; and reducing the polyphony of the program containing the samples to 'mono' (or one voice), which only allows the latest note event to sound, muting the previous one already playing.[3] The flux of these two parameter choices is often described as 'drum-mode' in music technology parlance (and products) because it has been typically relied upon to ensure realistic drum sequencing: drum sounds triggered on a keyboard, for instance, do not require a legato performance to ensure their full envelopes sound out (just as a drummer hitting a snare does not need to continue applying pressure with a stick for the snare to produce a full acoustic event, including any associated room reflections); while an open high-hat triggered would mute a closed high-hat already programmed, as a drummer would not produce both sounds simultaneously in a live performance (for this to work, samples can be given 'sounding' priority or, on the contrary, be set a 'mute targets'). And here lies the poetic oxymoron when 'drum-mode' is unleashed upon rich phonographic segments (instead of just individual drum sounds) captured within the sequences of sampling drum machines like the MPC: we are hearing powerful, percussive mobilisations of full phonographic events, interacting with (indeed interrupting) each other in highly syncopated, drum-like juxtaposition. A prime example of this rhythmic-motivic signature can be heard on Gang Starr's *Hard To Earn* (1994) album.

Textural signatures

In the first volume of *Perspectives on Music Production*, Matt Shelvock (2017, p. 170) discusses how the notion of hip-hop production—as "beat-making", composition, or music creation—is closely integrated with mixing practices and "that the lines between mixing and production are often blurred within this genre". As such, he highlights how hip-hop mix engineers may be closely involved with creative production decisions, and that clear stages between making and mixing a track may not be adhered to in hip-hop practice. Looking at this relationship from the perspective of the beat-maker, it is also true that the hip-hop producer assumes a plethora of mixing roles *while* creating a track. One of the defining features of the MPC workflow was the inclusion of mixing/processing functionality, which empowered beat-makers with sonic options traditionally reserved for the mixing stage. What's more, the MPC operating script caters for a flexible routing functionality that makes possible the insertion of effects at various stages (upon an individual sample, program, track, or the master). The resulting 'signal flow' outcomes create striking and unique sonic signatures, which would have required complex processes were they attempted within a DAW environment or via patch-bay routing in a hardware studio.

It is also important to discuss the limitations imposed by the internal effect processing capabilities of the early DSP chips on MPCs and note the aesthetic implications these may have had on the musical outputs of respective eras. The early chips limited the amount of effects that could be internally patched in at the same time (as insert effects or in an auxiliary configuration) to two, so for the sake of efficiency, practitioners would often share effects across a number of programs or tracks. For instance, a typical configuration would be to enable a master compressor for the whole internal MPC mix 'bus', while also sharing a reverb effect across numerous elements. This limitation would often result in a notion of sonic 'glue' or 'blend', an illusion constructed out of shared spatial, textural, and dynamic characteristics.[4]

Compression hereby deserves special mention, because its (ab)use by practitioners within the context of Hip Hop and the MPC environment has resulted in very particular stylisations (these have been exponentially expressed in the more experimental outputs of Flying Lotus, Madlib, J-Dilla, and Prefuse 73, as discussed by D'Errico (2015) and Hodgson (2011)). Although compression is a dynamic effect common to most virtual or physical mixing environments, its inclusion within the MPC operating script, the possibility to insert master compression on the stereo mix-bus (as part of the internal processing matrix), and the combined effect of the compression algorithm with that of the reduced bit-depth resolution of earlier models (and its emulation on later ones) lends it a unique quality in this context.

Furthermore, the physical input and output connectors on the MPC range, the analogue-to-digital and digital-to-analogue converters, and the resolution capabilities of various models, contribute to particular mixing affordances, sonic signatures, and lo-fi signifiers. The ability to produce a full musical piece on an MPC meant that it became possible for beat-makers to bypass the use of a computer sequencer or DAW altogether and bring a complete instrumental idea into the studio for mixing. Akai facilitated multiple outputs on most models of the hardware (either as default or as an optional expansion) allowing for the totality of an instrumental production to be output directly onto a hardware studio console. The sonic character of the physical I/O on MPCs (including the vinyl preamp inputs on most models) would thus be imprinted onto both incoming (sampled) sources, as well as outgoing multitracks, contributing to a recognisable timbral footprint. The sound of certain MPC models has become revered to such an extent that recent reincarnations of the technology feature emulations of earlier models (for example, the 'vintage mode' on the MPC Renaissance and MPC X, providing emulations of the MPC 60 and the MPC 3000). The reduced bit-rate (resolution) of earlier models is one of the notable variables impacting the audio character of hip-hop outputs from respective eras. This brings us to the present, and it is important to investigate the current footprint of these stylisations and their relationship to contemporary musical outputs in Hip Hop's trajectory.

Natural (sample) selection

> I give you bars, no microwave rap
> I can take it down South, but it's gon' be my version of trap …
> I don't hate the trap but give me that boom bap
> Yeah the 808 eating at the beats drill the 808
>
> *(Statik Selektah feat. The LOX & Mtume, 2017)*

The preceding lyrics are from Statik Selektah's 'But You Don't Hear Me Tho' (2017), highlighting the aesthetic friction between Trap—the prevailing and highly synthesised contemporary rap style—and the retrospective boom-bap sound. The track features many of the stylisations characteristic of Boom Bap, complete with prominent kick- and snare-drum sounds, highly swung programming, and chopped, soulful samples (albeit some recorded freshly by soul-funk band Mtume with additional horns from Utril Rhaburn). Statik Selektah is a contemporary beat-maker and DJ Premier protégé whose production duties are called upon when—according to XXLmag—"classic hip-hop at its finest" is required (Emmanuel, 2017). He represents a generation of producers and artists that *Pitchfork* describes—alongside rapper Your Old Droog—as "a nostalgic sonic wave currently being surfed by NYC contemporaries Roc Marciano, Action Bronson, and Joey Bada$$, rappers doing their best to embody the spirit of New York hip-hop without getting stuck in its past" (Ruiz, 2017); or a "lineage of gritty mafioso rap … deconstructed and mutated by Roc Marciano and The Alchemist in the 2000s, and polished by Griselda Records … in the 2010s" (Kearse, 2022). Meanwhile, the boom-bap term enjoys a revival in the context of artists' lyrics and as a common stylistic descriptor in album reviews, too. Indicatively, Ruiz (2017) describes Your Old Droog's single 'Bangladesh' (from album *Packs* (2017)) as "an ill Bansuri loop over a simple boom-bap drum beat", and HipHopDX reviews his track 'Help!' as "an unremittingly noisy blast of psychedelic boom-bap" (Leask, 2017).

If Your Old Droog's *Packs* merges sparseness and a sample-based aesthetic with noisy psychedelia, then another release, Apollo Brown and Planet Asia's album *Anchovies* (2017), takes the recipe down to the absolute rawest of materials (as does much of Roc Marciano's self-produced catalogue—for example, *RR2 – The Bitter Dose* (2018), *Marcielago* (2020), and *Mt. Marci* (2021)). Planet Asia and Roc Marciano are, in fact, often credited as pioneers of the minimal avantgarde in contemporary Boom Bap, having paved the way for other self-producing artists such as Marciano's collaborator and protégé Ka (for an indicative example, check *Descendants of Cain* (2020)). Perhaps it is a sign of genre maturity when process is reduced to its leanest, and the overarching simplicity of albums such as *Anchovies*, *RR2*, or *Descendants of Cain* exposes Boom Bap's DNA in a minimal production approach that does away with obvious drum reinforcement, coming full circle to the turntablist tradition of chopped, flipped, and rewound instrumentals (with rather extraneous amounts of vinyl noise). When asked about his minimal use of drums in his evolving production ethos, Marciano (cited in Byrne, 2018) explains:

> For me as an MC I enjoy the space. Sometimes the drums, you know, the program takes over the groove and doesn't allow me the same space. A track with no drums gives me the space to do more.

As an exception to the onomatopoeic boom-bap dogma, this approach reveals the mechanics underneath the beat: chopped instrumental samples programmed into reimagined sequences and chord progressions, where swung quantisation drives a highly rhythmic and hypnotic interaction between sourced phonographic material and the newly enforced temporal relationships of a sequencer. The customary boom-bap drum beats are either minimal or simply implied, but the rhythmic placement of the chopped samples expresses the very essence of the sample-based aesthetic. It is not a stretch to suggest that music producers may require the distance of a few decades to identify the raw essence of a genre, and be ready to expose its fundamental mechanics and source materials with such transparency. At the heart of the artform lies a beat-maker working with an instrumental formula that affords these sonic and temporal relationships to manifest: a formula originally inspired by the MPC.

Reflections

Although the sample-based modus operandi that defines the boom-bap sound is represented by a large number of contemporary releases, it has to be acknowledged that it is not the prevalent style of Hip Hop currently in the mainstream. Conversely, Trap's reign over the genre for a considerable number of years requires further examination at a time when electronic music forms are subject to exponential "trans-morphing" into numerous subgenres (Beer and Sandywell, 2005). Some of the developing stylisations coming out of Trap's evolution (and its related hybrids, Drill for example) have been contaminating old-school-inspired, sample-based beat-making for some time. The sonic evidence of this is audible in the longer, more pitched, and distorted 808 sub basses, now transcending their traditionally utilitarian role of merely supporting the isolated kick drums of Boom Bap, to provide dark, melodious, and prominent bass lines of their own.[5] Yet, it is not rare for forms that cross over into the mainstream to simultaneously undergo an aesthetic counter-reaction, a phenomenon expressed by underground purveyors tracing and practising the mechanics and stylisations of older subgenre forms. This is encapsulated by subcultures becoming consciously retrospective and evolving their stylisations according to their own code of aesthetic conduct, one that is slower than the pace dictated by commercial pressures (Thornton, 2001). The currently buzzing hip-hop underground certainly represents such a reaction, and the answer in artists' lyrics and producers' practices seems to point towards a reimagined Golden Age boom-bap recipe.

But a pragmatic challenge lies in the sourcing of raw materials (phonographic samples) necessary for the process to function. Boom Bap's dependence on the past is challenged by the legal context and finite pool of phonographic material available to producers in the decades prior to the birth of sample-based Hip Hop. As a result, sample-based producers are forced to source alternative content should they continue to engage in boom-bap practices. The latest incarnations of the MPC range (MPC X, Key, Live, One) appear to be acknowledging the methodological alternatives contemporary producers practise, retaining the interface, operating system, and workflow affordances that powered Golden Age aesthetics, while maximising the potential for recording new music directly into its interface and leveraging interaction with synthesised music forms (exemplified by direct inputs for live instruments, CV outputs for analogue synth control, pre-loaded Trap and EDM sound libraries, and 'controller' modes for working with a computer). If music makers are adamant about pursuing and evolving the sample-based artform—going as far as reverse-engineering original sampling content—then the instrument that has been fuelling sample-based divergences since 1988 may just be able to support this retrospective-futuristic oscillation that characterises so much of meta modern creative practice (Vermeulen and Van Den Akker, 2010). Yet, the stylisations it has afforded, and the workflow tendencies it has conditioned, are now part and parcel of producers' global vocabularies practised beyond the context of MPC technology and the confinements of the boom-bap aesthetic: the sample-based, syncopated lo-fi 'chop' may just have become the guitar riff of the microchip era.

Practical sonification

In this section, I highlight a number of beats from the accompanying album, which illustrate both key boom-bap signatures discussed throughout the chapter, as well as contemporary hybridisations owing to the trans-morphing contamination by way of the more synthetic or experimental subgenres, such as Trap and Glitch Hop. All of these productions, however, have been actualised by first creating and capturing the source material from scratch, mirroring some of the practices deconstructed in Chapter 1.

Reframing and layering

Track 'Rebluezin', as the name suggests, acts as thematic link to the previous chapters, deploying self-created multitrack masters from the blues pool of content (featuring upright piano, blues harp, acoustic drums, and electric bass), but reframing them into a sparser, darker, and more ambient soundscape (and more minor tonality). This has been achieved by chopping very short segments from a blues instrumental, pitching them down by about seven semitones, and rhythmically triggering new sequences from an MPC X's drum pads by hand (with the resulting blues sample program in monophonic, one-shot mode). The 'winning' motifs from the percussive 'jam' become the juxtaposed layers giving way to the A-B structure audible in the first eight bars of the production (the busier four-bar intro, and the more minimal four bars of the consecutive instrumental 'verse'). A few varisped blues vocal samples add emphasis at the end of select two-bar phrases throughout different sections of the track. A high level of swing quantisation has been enforced upon the instrumental sequences following the performance, making conscious use of the MPC's characteristic rhythmic signature. This also allows for a tightly matched 'reinforcement' from the added individual kick, snare, and high-hat drum hits, which adhere to the same swing percentage as the blues 'chops' (and pay sonic homage to producers 9th Wonder and The Alchemist in equal measure). The combined boom-bap structure is manifested in the syncopated layering of the added drum hits and blues chops, while the latter are allowed to 'sound off' until they are muted by the next segment triggered, revealing some of their inherent syncopation within the original instrumental performances. The triplet feel of the performances sampled aligns with the highly swung MPC programming, and becomes most exposed in the interaction between the blues harp and original snare drum in the last beat of bars six and eight (made even more emphatic by the proximity of the sampled snare to the next downbeat in the sequence). A pitched, detuned, and distorted 808 bass drum with long decay is melodically performed on the MPC's drum pads acting as a new bass line for the track. It mostly supports the programmed kick drums, but also plays 'off' them, at times landing on the odd snare hit, providing a passing note in the form of a lower fifth; elsewhere in a four-bar loop, mirroring the blues harp gestures; and whenever the opportunity presents itself, adding a swung sixteenth just before the next downbeat (in place, or alongside a double kick drum).

In order to sonically work with the 808 bass line, the blues samples have been further processed within the MPC environment. Using the onboard effects processors, the blues program is high-pass filtered, creating some space at the bottom of the frequency spectrum, with its upper bass- and higher mid-range accentuated to highlight particular aspects of the sample's instrumentation. Additionally, the blues program has been distorted through an amp emulator, attempting to match some of the distortion characteristics of the sub bass and add harmonics to the blues samples, somewhat 'aging' the instrumentation (hinting at complimentary media-based staging particularly for the often-exposed blues harp). Inspired by the sparseness of the blues chops' placement in the minimal sequences of the beat (the B sections), the samples are then sent to a rather long reverb emulation of a stadium space, as well as a psychedelic quarter delay emulator with high feedback settings. These ambient statements, in turn, have inspired additional instrumentation overdubbed progressively upon the developing production: namely, the harpsichord (spinet) performances, electric guitar licks, and the horn samples introduced at bar 34.

The Telecaster guitar and harpsichord motifs have been performed 'on top' of the looping sample-based structure and recorded directly into the MPC, before being chopped into shorter segments and re-sequenced. However, a number of signal flow and tracking strategies are observed to match the sonics of these later additions to the evolving sample-based sonic stage of the main beat. The Telecaster has been tracked through a valve guitar amp, with generous amounts of onboard spring reverb, using both close (dynamic) and farther (condenser) microphones to capture an ambient footprint. The Ray Manzarek/ Doors-inspired harpsichord performance has been tracked in the same recording space as the guitar amp, with the farther mic repositioned at some distance from the sounding board of the instrument. Both instrumental sources are run through a combination of hardware and software-emulated console pre-amps, consistent to the era-informed choices responsible for the recording and mixing of the earlier blues sample content. The harpsichord is then processed within the MPC environment, made to 'share' some of the ambient spaces and psychedelic delays staging the sampled elements of the beat. The horns introduced later in the arrangement come from the Afro-Cuban body of recordings, having been chopped, similarly processed, and pitched down rather radically to match the key and rhythm of the previous layers. Just like the blues multitrack sampled, the horn elements have been played out of D/A converters (isolated from their original mix), and straight into the A/D inputs of the MPC. Finally, sub-groups of drum, instrumental, and sampled elements are routed out of the separate outputs of the MPC X (via MPC 3000 and MPC 60 emulations) into a hardware summing mixer, taking advantage of harmonic colouration, parallel dynamics, multiple transformer stages, and bus compression in the analogue domain—the totality of the tracking, MPC, and post-production processing aiming at a coherent sonic 'whole' created out of the disparate elements, with recorded samples of vinyl crackle rhythmically overlaid throughout to provide an illusion of shared 'format' origins. Within the virtual domain of the MPC environment, most elements are gain-managed well within the 'legal' margins of its digital headroom, with the exception of the added kick drum, which is purposefully 'clipped into the red' of an MPC 60 converter emulator, to assume the characteristically *crunchy sound* of old-school Boom Bap.

Chopping for 'hooks'

The percussive reimagining and appropriation of multitrack material from the different source eras/styles is evident on most tracks in the accompanying album, with the chopping workflow thus far described situated at the heart of what powers the underlying sample-based utterances. Tracks such as 'Covert', 'Rijeka', 'Kalimbap', 'Born To Death', 'Cycles', and 'Boom Bag', for example, deploy the majority of chopping and processing techniques highlighted for 'Rebluezin': jamming with sonic objects from a wide range of self-made phonographic scenarios in pursuit of sample-based 'hooks'. Respectively, 'Covert' is built around drum-pad performances over improvised phrases recorded with the Coventry Cathedral pipe organ, experimentally multi-miked around the space to capture both its direct sound from the pipes as well as its spatial footprint. Bars 13–20 reveal the motifs created with the isolated organ chops, which have inspired the remainder of the beat construction as well as the additional layers in the busier sections. It is worth noting that the sub frequencies audible in bars 17–20 are a result of chopping sections of the organ performance deploying the lowest organ pipes, which in turn dictated some of the added 808/sub musings. Further layers include a sliced drum break, vocal chops resembling vinyl scratches from the alt-rock/hardcore body of source material, and processed windchimes recorded as Foley using a phone. These same chimes become one of the main raw ingredients for 'Rijeka', a track that began as a beat-making exercise initially limited to Foley recordings captured in the city that lends the track its name (specifically, windchimes, church bells, and voices). The track also features the customary programmed (kick, snare, and high-hat) drum hits, sliced drum breaks, and 808 sub support, while the outro reveals an alternative reimagining of the Coventry Cathedral organ improvisations, brought in to assist the beat's climax in the later sections.

Trans-morphing

'Kalimbap' is a rather minimal affair built out of a juxtaposition of processed kalimba performances, run through hardware pedals, then chopped, re-sequenced, and made to interact in a kind of call-and-response fashion with samples from the classical/orchestral body of source material. The samples in question are mixes of an overdubbed harpsichord and operatic vocal improvisation, the latter in a made-up language. What is key about this production is that although the sample-based motifs, drum hit choices, and 808 (sub) reinforcement adhere to a boom-bap modus operandi, the more avantgarde kalimba-vocal interactions—particularly the glitchy stutters—have opened up the door to programming that references trap/drill stylisations. This is most evident in the high-hat patterns featuring dynamic thirty-second and twenty-fourth rolls (created using the MPC's note-repeat function, but taking advantage of the tactile pressure-sensitivity of the drum pads), as well as the legato, distorted 808 sub bass contributing to the track's bassline. 'Born To Death'—built around a minimalist, lyrically driven reframing of material from the art-rock body of source work—offers similar divergences, whilst adhering to the slower tempo of many a contemporary boom-bap-based hybridisation (some musicologists argue that Trap, for example, can be felt in two tempos: that of the main beat, and double that, courtesy of the high-hat programming—see Bennett (2020) for more). As such, 'Kalimbap' and 'Born To Death' are two of the tracks associated with this chapter that offer discreet degrees of 'trans-morphing' between the boom-bap aesthetic and the more synthetic hip-hop hybridisations.

Revelations

'Cycles' owes its A-B sound and structure to a more orthodox boom-bap reimagining of a psychedelic soul multitrack, created by mixing (Rhodes) electric piano, electric bass, electric guitar, and harpsichord performances, then chopping the master in real time during sampling, using the MPC's drum pads. Bars 1–4 and 13–16 reveal the two hook 'foundations' driving the beat, which are layered against drum hits, sliced drum breaks, and 808 programming for the remainder of the sections. As with many productions in the genre, breakdowns, intros, and outros such as these are strategically placed to reveal some of the underlying source's origins, before becoming reframed in the reimagined structure (see Chapter 4 for more on this process, known as 'flipping' in hip-hop parlance). 'Boom Bag'—a beat borne out of sampling an impromptu, overdubbed jam with a musical colleague on Greek baglama, acoustic guitar, percussion, Rhodes, harpsichord, and electric bass—makes such a revelation in its breakdown section, exposing a rather linear sequence from the source's mix/master. What's different about the workflow here, in comparison to preceding beats, is that the track's sections before and after the breakdown benefit from a kind of 'remixing' freedom, sampling and chopping isolated elements from the multitrack that contributes to the end stereo mix. In order to preserve sonic coherence, two strategies are deployed: the deconstructed mix elements retain all of their mix colouration, as well as their textural and spatial processing (their sonic DNA, so to speak, as part of the larger mix/master organism) before getting sampled; and various 'gluing' strategies are pursued within and via the MPC environment's signal flow and internal processing architectures (physical sampling via the converters, shared spatial effects, and shared MPC 60 and 3000 emulations, before finally routing the mix out of the MPC's D/A outputs and into the analogue summing chain).[6]

Hybridisation

Finally, tracks 'City', 'Kaishaku', and 'Skinwalkers' divert further into synthetic/trap-informed hybridisation (and the glitch-hop avantgarde), whilst keeping true to the boom-bap essence of the sound-object manipulations that power their founding structures. 'City' takes advantage of a prog-rock experiment (featuring multiple electric and acoustic guitar layers, analogue synthesisers, organs, Rhodes electric piano, electric bass, percussion, and drums) mastered through a(n iZotope) vinyl player emulator. The vinyl emulation software has been manipulated in real time—like a record player—during the sampling process, which results in the warping and stop/start effects audible in the production (particularly in the more exposed intro/outro/breakdown sections). Much of the workflow and strategies described for the previous body of tracks are also deployed in the making of this beat, with the exception of the more synthetic drum textures, here courtesy of a Behringer RD-8 drum machine (a modern hardware emulation of a Roland TR-808). The drum machine's sounds have been modulated on the RD-8, but then sampled into the MPC in order to benefit from its signal-flow colourations and quantisation characteristics. A hybrid of 808-style drum textures and sampled (acoustic) drums powers 'Kaishaku' and 'Skinwalkers'—reimaginings of heavy multitracks from the alt-rock body of work. The structure of 'Kaishaku' starts with a subtle introduction of very short, effected chops from the original master, before progressively revealing slowed-down but rather intact sections from the source in the chorus and outro structures. 'Skinwalkers', equally, deploys a plethora of short segments from the source master (recorded to and manipulated via a modified cassette player) to juxtapose overlaid performances of heavily processed chops using the MPC's drum pads, sequencer functionality, and physical controllers to perform the dub-like effect automation—the chaotic, deconstructed narrative is told through the dense layering that evolves from the hip-hop grooves of the early sections, through to the four-to-the-floor industrial rhythms of the climax.

For all of these beats, the source samples have been purposefully 'aged' in the sonic domain through the deployment of vintage workflows, physical equipment, recording formats/media, or emulations thereof. The following chapter deals with some of the conceptual issues in this undertaking, exploring the philosophical dimensions of consciously invoking past sonics as part of contemporary beat-making practice.

Recommended chapter playlist
(in order of appearance in the text)

'Rebluezin'
'Covert'
'Rijeka'
'Kalimbap'
'Born To Death'
'Cycles'
'Boom Bag'
'City'
'Kaishaku'
'Skinwalkers'

Notes

1 Clarke expands Gibson's (1966/1983, p. 285) coining of the term *affordance* from "a substitute for *values* ... what things furnish ... (w)hat they *afford* the observer ... (which) depends on their properties" into a concept that takes into account the social interdependence between objective properties and the nature of human users/listeners; according to Clarke (2005, p. 38), "affordances are primarily understood as the *action* consequences of encountering perceptual information in the world".
2 Indicative examples include *Bare Face Robbery* (Dirt Platoon, 2015), *Mr. Wonderful* (Bronson, 2015), *B4.DA.$$* (Bada$$, 2015), *Super Hero* (Kool Keith feat. MF Doom, 2016), *That's Hip Hop* (Ortiz, 2016), *The Ghost of Living* (Vic Spencer & Big Ghost LTD, 2016), *Ode to Gang Starr* (Brown, 2017), *The Good Book*, Vol. 2 (The Alchemist & Budgie, 2017), *Underground with Commercial Appeal* (Fokis, 2017), *In Celebration of Us* (Skyzoo, 2018), *Sincerely, Detroit* (Brown, 2019), *WWCD* (Griselda, 2019), *No One Mourns the Wicked* (Conway x Big Ghost LTD, 2020), *Amongst Wolves* (SmooVth & Giallo Point, 2020), and *If You Know You Know* (Elcamino, 2021).
3 Program polyphony hereby refers to MIDI voices, not the true nature of musical polyphony that may be contained within a sampled sonic object (the MPC relies on MIDI data to action and recall triggering of audio segments).
4 In a study that focuses primarily on compression in mastering, Moore (2021, p. 58) provides a useful, if narrow, definition of 'glue' as a characteristic that "creates a cohesiveness to program material", and which "may impart subtle distortion, colouration and rhythmic movement". Citing Cousins and Hepworth-Sawyer (2013, p. 74), he also preempts it with a wider understanding as "a by-product of gain control, making the track sound like a whole entity rather than its individual parts".
5 Great examples of this developing trap/drill signature audible in boom-bap productions are tracks such as 'For Sale' (from *T.O.N.Y. 2* (Pounds448, 2022)) produced by Chukk James, and 'Hear Me Clearly' (from *It's Almost Dry* (Pusha, 2022)) produced by BoogzDaBeast, Luca Starz, TheMyind, and Kanye West (for more on the relationship between synthetic and sample-based aesthetics throughout the history of Hip Hop, see Exarchos, 2020).
6 The creative implications and rationale of this form of remixing workflow upon sampling (isolated) multitrack material is discussed in detail in Chapter 6.

Bibliodiscography

Aikin, J. (1997) 'Samplers Rule', *Keyboard*, October, p. 47.

Anderton, C. (1987) *SP-1200 owner's manual*. Scotts Valley, CA: E-MU Systems.

Avgousti, A. (2009) *Beatmaking on the MPC2000XL*. 3rd edn. MPC-Samples.com.

Bada$$, J. (2015) *B4.DA.$ [Digital Release, Album]*. US: Pro Era, Cinematic Music Group.

Beer, D. and Sandywell, B. (2005) 'Stylistic morphing: Notes on the digitisation of contemporary music culture', *Convergence: The International Journal of Research into New Media Technologies*, 11(4), pp. 106–121.

Bennett, J. (2020) 'What Makes Trap… Trap? Our Resident Musicologist Unpacks the Era-Defining Sound', 8 June. Available at: https://tidal.com/magazine/article/what-makes-trap-trap/1-72722 (Accessed: 4 October 2022).

Bronson, A. (2015) *Mr. Wonderful* [CD, Album]. US: Atlantic/Vice Records.

Brown, A. (2019) *Sincerely, Detroit* [CD, Album]. US: Mello Music Group.

Brown, A. and Asia, P. (2017) *Anchovies* [Stream, Album]. US: Mello Music Group.

Brown, S. (2017) *Ode To Gang Starr* [Digital Release, Single]. US: Below System Records.Byrne, M. (2018) '"I'm Always Trying to Keep Pushing Myself": An Interview with Roc Marciano', *Passion of the Weiss*, 29 March. Available at: https://www.passionweiss.com/2018/03/29/roc-marciano-interview/ (Accessed: 7 September 2021).

Chang, J. (2007) *Can't stop won't stop: A history of the hip-hop generation*. Reading, PA: St. Martin's Press.

Clarke, E.F. (2005) *Ways of listening: An ecological approach to the perception of musical meaning*. Oxford: Oxford University Press.

Conway x Big Ghost LTD (2020) *No One Mourns the Wicked* [CD, Album]. US: Gourmet Deluxxx.

Cousins, M. and Hepworth-Sawyer, R. (2013) *Practical mastering: A guide to mastering in the modern studio*. New York: Focal Press.

D'Errico, M. (2015) 'Off the grid: Instrumental Hip-Hop and experimentation after the Golden Age', in J.A. Williams (ed.) *The Cambridge companion to Hip-Hop*. Cambridge: Cambridge University Press, pp. 280–291.

Dery, M. (1990) 'Hank Shocklee: "Bomb squad" leader declares war on music', *Keyboard*, pp. 82–83, 96.

Dirt Platoon (2015) *Bare Face Robbery* [Stream, Album]. US: Effiscienz.

Elcamino (2021) *If You Know You Know* [Digital Release, Album]. US: Anti Gun Violence Co.

Emmanuel, C.M. (2017) *The Lox and Mtume Show Off on Statik Selektah's 'But You Don't Hear Me Tho'*, *XXL Mag*. Available at: https://www.xxlmag.com/the-lox-and-mtume-show-off-on-statik-selektahs-but-you-dont-hear-me-tho/ (Accessed: 19 September 2017).

Eric B. and Rakim (1987) *Paid In Full* [CD, Album]. Europe: 4th & Broadway, Island Records.

Exarchos, M. (2020) 'Synth sonics as stylistic signifiers in sample-based Hip-Hop: Synthetic aesthetics from "Old-School" to Trap', in N. Wilson (ed.) *Interpreting the synthesizer: Meaning through sonics*. Newcastle upon Tyne: Cambridge Scholars Publishing, pp. 36–69.

Fokis (2017) *Underground with Commercial Appeal* [Digital Release, Album]. US: Loyalty Digital Corp.

Gallant, M. and Fox, D. (2016) *Creating Hip Hop: 10 Questions with Legendary Producer Just Blaze*, *Universal Audio*. Available at: https://www.uaudio.com/blog/creating-hip-hop-with-just-blaze/ (Accessed: 19 September 2017).

Gang Starr (1989) *No More Mr Nice Guy* [CD, Album]. US: Wild Pitch Records.

Gang Starr (1992) *Daily Operation* [CD, Album]. UK & Europe: Cooltempo.

Gang Starr (1994) *Hard to Earn* [CD, Album]. US: Chrysalis.

George, N. (2005) *Hip Hop America*. London: Penguin Books.

Gibson, J.J. (1983) *The senses considered as perceptual systems*. Westport, CT: Greenwood Press.

Grandmaster Flash (1981) *The Adventures of Grandmaster Flash on the Wheels of Steel* [Vinyl LP]. US: Sugar Hill Records.

Griselda (2019) *WWCD* [Digital Release, Album]. US: Griselda Records, Shady Records.

Harkins, P. (2008) 'Transmission loss and found: The sampler as compositional tool', *Journal on the Art of Record Production*, 4.

Harkins, P. (2010) 'Appropriation, additive approaches and accidents: The sampler as compositional tool and recording dislocation', *Journal of the International Association for the Study of Popular Music*, 1(2), pp. 1–19.

Hodgson, J. (2011) 'Lateral dynamics processing in experimental Hip Hop: Flying Lotus, Madlib, Oh No, J-Dilla and Prefuse 73', *Journal on the Art of Record Production*, 5.

Howard, D.N. (2004) *Sonic Alchemy: Visionary music producers and their maverick recordings*. Milwaukee: Hal Leonard Corporation.

Jaylib (2003) *Champion Sound* [Stream, Album]. US: Stones Throw Records.
Jay-Z (2017) *The Story of O.J., 4:44* [Digital Release, Single]. Roc Nation.
Ka (2020) *Descendants of Cain* [CD, Album]. US: Iron Works Records.
Kajikawa, L. (2015a) 'Historicizing rap music's greatest year', in J.A. Williams (ed.) *The Cambridge companion to Hip-Hop*. Cambridge: Cambridge University Press, pp. 301–314.
Kajikawa, L. (2015b) *Sounding race in rap songs*. Oakland: University of California Press.
Katz, M. (2012) *Groove music: The art and culture of the Hip-Hop DJ*. New York: Oxford University Press.
Kearse, S. (2022) *Dollar Menu 4, Pitchfork*. Available at: https://pitchfork.com/reviews/albums/mach-hommy-dollar-menu-4/ (Accessed: 6 October 2022).
Kool Keith feat. MF Doom (2016) *Super Hero* [Digital Release, Single]. US: Mello Music Group.
KRS-One (1993) *Return of the Boom Bap* [CD, Album]. US: Jive.
Kulkarni, N. (2015) *The periodic table of Hip Hop*. London: Random House.
Leask, H. (2017) *Review: Your Old Droog Steps His Rap Game Up On 'Packs', HipHopDX*. Available at: https://hiphopdx.com/reviews/id.2917/title.review-your-old-droog-steps-his-rap-game-up-on-packs (Accessed: 19 September 2017).
Leight, E. (2017) *'4:44' Producer No I.D. Talks Pushing Jay-Z, Creating '500 Ideas', Rolling Stone*. Available at: https://www.rollingstone.com/music/music-features/444-producer-no-i-d-talks-pushing-jay-z-creating-500-ideas-253045/ (Accessed: 6 October 2017).
Linn, R. (1989) *MPC60 Midi Production Centre: Operator's manual*. Yokohama: Akai Electric Co.
Marciano, R. (2018) *RR2- The Bitter Dose* [CD, Album]. US: Fat Beats.
Marciano, R. (2020) *Marcielago* [CD, Album]. US: Fat Beats Distribution.
Marciano, R. (2021) *Mt. Marci* [CD, Album]. US: Marci Enterprises, Art That Kills.
McIntyre, P. (2015) 'Tradition and innovation in creative studio practice: The use of older gear, processes and ideas in conjunction with digital technologies', *Journal on the Art of Record Production*, 9.
Mlynar, P. (2013) *In Search of Boom Bap, Redbullmusicacademy*. Available at: https://daily.redbullmusicacademy.com/2013/11/in-search-of-boom-bap (Accessed: 19 September 2017).
Moore, A. (2021) 'Towards a definition of compression glue in mastering', in J.-P. Braddock et al. (eds) *Mastering in music*. London: Routledge (Perspectives on Music Production), pp. 44–59.
Morey, J.E. and McIntrye, P. (2014) 'The creative studio practice of contemporary dance music sampling composers', *Dancecult: Journal of Electronic Dance Music Culture*, 1(6), pp. 41–60.
Ortiz, J. (2016) *That's Hip Hop* [CD, Album]. US: Deranged Music Inc.
Perron, M. (1994) 'Checking tempo stability of MIDI sequencers', in *Proceedings of the 97th Audio Engineering Society Convention*. San Francisco.
Pounds448 (2022) *T.O.N.Y. 2* [Digital Release, Album]. US: 1132536 Records DK.
Pusha T (2022) *It's Almost Dry* [CD, Album]. US: Def Jam Recordings, Getting Out Our Dreams.
Ratcliffe, R. (2014) 'A proposed typology of sampled material within electronic dance music', *Dancecult: Journal of Electronic Dance Music Culture*, 6(1), pp. 97–122.
Rodgers, T. (2003) 'On the process and aesthetics of sampling in electronic music production', *Organised Sound*, 8(3), pp. 313–320.
Rose, T. (1994) *Black noise: Rap music and black culture in contemporary America*. Hanover, NH: University Press of New England (Music/Culture).
Ruiz, M.I. (2017) *Your Old Droog: Packs, Pitchfork*. Available at: https://pitchfork.com/reviews/albums/22988-packs/ (Accessed: 19 September 2017).
Scarth, G. and Linn, R. (2013) *Roger Linn on Swing, Groove & the Magic of the MPC's Timing, Attack Magazine*. Available at: https://www.attackmagazine.com/features/interview/roger-linn-swing-groove-magic-mpc-timing/ (Accessed: 10 May 2018).
Schloss, J.G. (2014) *Making beats: The art of sample-based Hip-Hop*. Middletown, CT: Wesleyan University Press (Music/Culture).
Serrano, S., Torres, A. and Ice-T (2015) *The rap year book: The most important rap song from every year since 1979, discussed, debated, and deconstructed*. New York: Abrams Image.
Sewell, A. (2013) *A typology of sampling in Hip-Hop*. Unpublished PhD thesis. Indiana University.

Shelvock, M. (2017) 'Groove and the grid: Mixing contemporary Hip Hop', in R. Hepworth-Sawyer and J. Hodgson (eds) *Mixing music*. New York: Routledge (Perspectives on Music Production), pp. 190–207.

Skyzoo (2018) *In Celebration of Us* [CD, Album]. US: Empire, First Generation Rich, Inc.

SmooVth & Giallo Point (2020) *Amongst Wolves* [Vinyl LP]. Denmark: Copenhagen Crates.

Statik Selektah feat. The LOX & Mtume (2017) *But You Don't Hear Me Tho* [Digital Release, Single]. US: Showoff Records/Duck Down Music Inc.

Sugarhill Gang (1979) *Rapper's Delight* [Vinyl LP]. US: Sugar Hill Records.

Swiboda, M. (2014) 'When beats meet critique: Documenting Hip-Hop sampling as critical practice', *Critical Studies in Improvisation*, 10(1), pp. 1–11.

T La Rock and Jazzy Jay (1984) *It's Yours* [Vinyl LP]. US: Partytime Records, Def Jam Recordings.

The Alchemist & Budgie (2017) *The Good Book, Vol. 2* [Stream, Album]. US: ALC.

Thornton, S. (2001) *Club cultures: music, media and subcultural capital*. Reprint. Cambridge: Polity Press.

Tingen, P. (2007) *DJ Premier, SoundOnSound*. Available at: https://www.soundonsound.com/people/dj-premier (Accessed: 19 September 2017).

Toop, D. (2000) *Rap attack 3: African Rap to global Hip Hop*. 3rd edn. London: Serpent's Tail.

Vermeulen, T. and Van Den Akker, R. (2010) 'Notes on metamodernism', *Journal of Aesthetics & Culture*, 2(1), pp. 56–77.

Vic Spencer & Big Ghost LTD (2016) *The Ghost Of Living* [Digital Release, Album]. US: Perpetual Rebel.

Wang, O. (2014) 'Hear the drum machine get wicked', *Journal of popular music studies*, 26(2–3), pp. 220–225.

Weingarten, C.R. (2010) *It takes a nation of millions to hold us back*. New York: Continuum (33 1/3).

Williams, J.A. (2010) *Musical borrowing in Hip-Hop music: Theoretical frameworks and case studies*. Unpublished PhD thesis. University of Nottingham.

Your Old Droog (2017) *Packs* [CD, Album]. US: Fat Beats.

PART 2

(Reverse) Engineering

3

SONIC PASTS IN HIP HOP'S FUTURE

As seen in Chapter 1, much has been written in the literature about sampling as composition, the legality and ethics of sampling, and sampling as a driver of stylistic authenticity in Hip Hop. Several scholars have also dealt with the historicity of samples from a number of perspectives. In *Black noise*, Rose (1994, p. 79) effectively demonstrates how hip-hop producers consciously quote from a musical past they resonate with as a form of cultural association: "For the most part, sampling, not unlike versioning practices in Caribbean musics, is about paying homage, an invocation of another's voice to help you say what you want to say". In *Making beats*, Schloss (2014) reveals a complex ethical code shared by 1990s boom-bap (Golden Age) hip-hop practitioners, with strict rules about the periods, records, and particular content that may or may not be sampled. It is worth stating that sampling ethics in Hip Hop have much more to do with adherence to (sub)cultural codes of practice than copyright law.

On the other hand, Simon Reynolds (2011, pp. 314–315) states in *Retromania*:

> It's curious that almost all the intellectual effort expanded on the subject of sampling has been in its defence … A Marxist analysis of sampling might conceivably see it as the purest form of exploiting the labour of others.

Despite taking a critical stance towards the politics, ethics, and economics of sampling, Reynolds here highlights a number of important problems. From the perspective of a critic who does not necessarily enjoy sample-based artefacts—and therefore self-admittedly fails to understand the popularity of musics such as Hip Hop—Reynolds (2011, pp. 313–314), however, focuses our attention on the multi-dimensionality inherent in the complex phenomenon of 'recordings within recordings':

> Recording is pretty freaky, then, if you think about it. But sampling doubles its inherent supernaturalism. Woven out of looped moments that are like portals to far-flung times and places, the sample collage creates a musical event that never happened … Sampling involves using recordings to make new recordings; it's the musical art of ghost co-ordination and ghost arrangement.

DOI: 10.4324/9781003027430-7

With this observation, Reynolds provides an eloquent description of the sample-based phonographic *condition*. As such, the associated problems become key considerations for any practice attempting a process of reverse-engineering: how could an awareness of this exponential or 'supernatural' multi-dimensionality inform alternative practices that pursue a sample-based aesthetic? Morey and McIntyre (2014) criticise Reynolds for ignoring the contribution of the sampling composer in this position, thus adding to the complexity of the creative equation. The tension between their position and Reynolds's is perhaps a symptom of a larger philosophical problem: in attempting to serve a sample-based aesthetic through re-construction, a practitioner comes face-to-face with the irony between materiality and cultural referencing. Does a short sound contain history, 'style', a unique sonic signature? When is this historicity motivic, i.e. relating to melody, rhythm, and performance? Conversely—when not—what are the inherent sonic manifestations that infuse phonographic 'resonance' to a minute sonic segment? Can these be recreated? Zak (2001, pp. 195–197) concludes in *The poetics of rock*:

> The overall resonant frame amplifies, as it were, the smallest nuances with which records are filled... [Record collections] represent historical documents and instruments of instruction that provide both ground and atmosphere ... Collectively, records present an image of a cultural practice whose conceptual coherence is assured ... by the shared perception that its works possess the power of resonance.

Zak here supports the idea that cultural resonance can be embedded within the sonic grain of a record and consequently hints at a matrix of interrelationships situated between phonographic artefacts of different eras. Chapter 1 argued that Hip Hop is inherently inter-stylistic, its process resulting in new musical forms borne out of the manipulation of past ones; while at the same time morphing into numerous sub-genres, due to the speed and power of the dissemination and interaction afforded by digital technology. In their article on 'Stylistic morphing', Beer and Sandywell (2005, p. 119) theorise convincingly on this phenomenon:

> It seems that there is no such thing as genre ... Under further scrutiny canons prove to be complex configured collections of stylistic signifiers traversing cultural fields and interwoven with cultural objects. Against this paradoxical conclusion we suggest that genre is more than a technical or theoretical term. It is also a practitioner's term invoked in the recognition, consumption, and production of musical performances.

The above is a useful description of the creative flux facilitated by digital tools from the perspective of practitioners, and it has the potential to inform the theoretical framework behind a reverse-engineering process. Although the majority of rap practitioners may be *reacting* creatively to the cultural and legal context surrounding them—rather than first theorising about it—a number of telling positions towards sampling, characteristic of different rap eras, shed light onto the spectrum of creative possibility. The performative tradition of isolating, repeating, elongating, and juxtaposing sections from phonographic records on turntables by DJ pioneers such as Kool Herc, Grandmaster Flash, and Afrika Bambaataa signifies the DNA of the artform. But as we saw in the previous chapter, this was a while before *The Adventures of Grandmaster Flash on the Wheels Of Steel* (1981) provided a phonographic 'exception', committing the performative tradition of 'turntablism' to record. The importance of the release is that it carries an early manifestation of what Reynolds (2011, pp. 313–314) describes as the process of "using recordings to make new recordings". At this point in hip-hop history though, it had only been through turntable performance that the 'citation' and manipulation of previously released records could be committed phonographically; which explains why the majority of non-live Old School rap releases utilised synthesisers and drum machines to provide the electro-rap instrumentals that functioned as an alternative to live performance.[1]

Fast-forwarding to the mid-to-late 1980s—and the more widespread availability of affordable sampling technology—a number of seminal releases leveraged sample-based composition and arrangement, taking advantage of the record industry's initial inertia in (legally) reacting to the creative manifestations afforded by sampling. Records such as *It Takes a Nation of Million to Hold Us Back* (Public Enemy, 1988), *Fear of a Black Planet* (Public Enemy, 1990), and *Paul's Boutique* (Beastie Boys, 1989) are rumoured to contain hundreds of samples of previously released phonographic content, signifying maximal masterpieces of the sample-based artform that are often compared to a kind of rap musique concrète (LeRoy, 2006; Weingarten, 2010; Sewell, 2013). And yet, by 1991, the shift in the legal landscape kick-started a case of legal necessity becoming the driver of sonic innovation.

One such notable reaction can be observed in Dr. Dre's infamous flavour of interpolation. Dre's initial success with N.W.A. afforded him access to an era of musicians he revered—musicians that he could invite into the studio to (re)play elements of their own records, facilitating his sampling endeavours. Using the original players, instruments, and technology enabled the acquisition of authentic sonics from a different era, but without the need to pay high sampling premiums to record companies (holding the mechanical copyright). His heavy dependence on P-funk sonics was so impactful that it birthed a geographical divergence in Hip Hop known as West-Coast Rap (or G-funk); one that was diametrically opposed to New York's East Coast aesthetic, remaining synthesiser-heavy (and often sample-averse in a mechanical sense). Further reactions to the legal landscape, and the decreasing creative opportunities for phonographic sampling, can be summarised in three overarching approaches:

1. Live performed Hip Hop;
2. The construction of content replacing samples referenced/used in hip-hop production; and
3. The creation of original but era-referential content that can act as new sampling material.

A number of practitioner case studies are discussed below, exemplifying these practices.

Case studies

Live Hip Hop

The Roots are perhaps the most famous case in point for a predominantly live-performing (and recording) hip-hop band; and whilst they remain conscious of the aesthetic compromises resulting from not always directly interacting with sampling technology, the delineation of their outputs from proto-rap/live-based instrumentals are a result of exhaustive research on sample-based utterances and sonics, which are manifested in their performance practices, choice of instruments, and studio approaches (Marshall, 2006). Yet what sparks stylistic criticisms directed at them by the hip-hop community at large is the fact that their live sonics and musicality have not always sufficiently interacted with sampling processes and their resulting artefacts. In the words of their own manager, The Roots' debut album, *Organix* (1993), has been characterised as: "swag deficient, lacking the grit of *sample, microchip*, and identifiable urban narrative that, to this day, define the genre" (Thompson and Greenman, 2013, p. 101, my emphasis).

Creating sample-replacement content

J.U.S.T.I.C.E. League on the other hand is a production duo responsible for a plethora of contemporary rap hits (for artists such as Rick Ross, Gucci Mane, Drake, and Lil Wayne) who deploy methods that lie somewhere between interpolation and a convincing re-interpretation (and then manipulation) of referenced samples. In an interview with HotNewHipHop.com they shed light on the specifics of their process:

> Ok, we have a guitar – what kind of guitar was it? What was the pre-amp? What was the amp? What was the board that it was being recorded to? What kind of tape was it being recorded to? What kind of room was it in?
>
> *(Law, 2016)*

Law (2016) asserts that once they have "all the information available about the original sample, they begin ... recreating every aspect ... down to the kind of room it was recorded in". J.U.S.T.I.C.E. League's process reveals the importance of the sonic variables that lend a sample its particular 'aura'. Their meticulous re-engineering attempts to infuse convincing (vintage) sonics onto their referential, yet newly recorded, source content.

Creating new content for sampling

There are also increasing contemporary cases where practitioners create content infused with referential—stylistic and historical—attributes, but without direct semblances to previously released compositions. Producer Frank Dukes meticulously records sonically referential, but musically original, vintage-sounding material, to facilitate his sample-based production process. When this level of reverse-engineering is applied to completely original creations, the potential exists for musical innovation that, nevertheless, adheres to the sonic requirements of the sample-based aesthetic. Interviewed in Fader magazine, Adam Feeney a.k.a. Frank Dukes explains in his own words: "I'm still using that traditional approach, but trying to create music that's completely forward-thinking and pushing some sort of boundary" (cited in Whalen, 2016). Expanding on this approach in relation to Dukes's production of 'Real Friends', from *The Life of Pablo* (West, 2016), Whalen (2016) explains:

> [T]he song's "sample," [is] a delicate piano loop that sounds like it's lifted from a dusty jazz record, but that Dukes found without having to dig for anything, because he made it himself ... Manipulating his own compositions like they were somebody else's is a technique that has brought Feeney—an avowed crate-digger turned

self-taught multi-instrumentalist—from relative obscurity to a go-to producer for the industry's elite.

Theorising

Aesthetic problem #1: The function of nostalgia

In all the practitioner approaches described above lies a conscious approach to navigate the legal landscape safely, whilst establishing links with the past, either through motivic referencing (Dre's interpolation) or via sonic referencing (The Roots in their instrumental/studio choices; Frank Dukes and J.U.S.T.I.C.E. League in their painstaking recreation of vintage sonic signatures). Hip Hop is approaching almost five decades of existence at the time of writing (*44th Anniversary of the Birth of Hip Hop*, 2017), so could its obsession with the past be regarded as a metaphor for approaching a stylistic middle-life crisis? Or is this form of sonic nostalgia a wider symptom in popular music, as Reynolds claims, which becomes exponential in a form of music that owes its very inception, architecture, and DNA to previous music forms? The website Metamodernism.com has published the following criticism on Reynolds's take on the phenomenon:

> Simon Reynolds states that popular (music) culture is suffering from retromania, an incurable addiction to its own past ...his analysis is based on a nineteenth century—and therefore very modern—notion of 'authenticity'. It makes himself a symptom of that which he criticizes: retromania.
>
> *(Van Poecke, 2014)*

Perhaps a metamodern predisposition is not essential for the criticism to stand: the problem with a mono-dimensional diagnosis of an aesthetic 'fault' (in this case, solely attributed to nostalgia) is that it is using the symptom as both diagnosis *and* condition. From antique hunters through to fashion designers and phonographic 'crate-diggers', it appears that a certain distance from the past allows the human mind the benefit of retrospective appreciation. But to avoid oversimplification, Hip Hop is a complex phenomenon that deserves more thorough analysis. Socio-economic and technological factors are entangled in its history, development, and sonic genetics, so nostalgia alone appears an easy escape notion, distracting from a meaningful investigation of the conditions shaping this more complex sub-cultural phenomenon.

In *Can't stop won't stop*, Chang (2007, p. 13) explains that social engineering, Kool Herc's Jamaican-derived sound-system mentality, the withdrawal of funding for instrumental musicianship in New York schools, and a technically-trained but unemployed young generation became the conditions for Hip Hop's 'big bang'. As a result, sample-based Hip Hop was borne out of improbable factors colliding and, as a result, old funk breaks became the instrumental bed for a generation that needed to dance, rap, come together, party, or rebel. From this point onwards, the DJ-as-performer had begun 'jamming' with musicians from the past, reacting to their utterances, interacting with their recorded performances, collaborating (in non-real-time), and manipulating their recordings live (just like King Tubby had previously done with dub multitracks in a recording studio environment); a trait that has been reproduced by sampling producers ever since via their interaction with (affordable) sampling technology.

It is not a stretch to consider that performing with turntables became a solitary alternative to improvising with a band—only, one recorded in the past—for a generation that was largely deprived of instrumental tuition and opportunity. Fast-forwarding to the condition of the current bedroom producer, one can observe a parallel in the solitary state of collaboration with the past: a plethora of historic audio segments residing in the hard disks, memory banks, and sampler pads of a contemporary hip-hop studio setup—providing the 'live-musician' resonances for a solitary performer/writer/producer to interact with. As a result, the sample-based hip-hop process could be described as a *jam across time* with pre-recorded musicians from the past, afforded by digital sampling technology (and initially turntables): in other words, a 'hip-hop time machine'. This can be represented schematically with the 'equation' in Figure 3.1.

The end result may sound nostalgic because of its obsession with the past, but it is really a manifestation of an inherent genetic trait that defines its very function and aesthetic. The sample-based condition has occupied such a large ratio of hip-hop outputs in its near-five-decade-long lifetime that it has elevated and celebrated the morphing, synthesis, and interaction of old and new music to the forefront of its modus operandi. Of course, Hip Hop has had an undeniable effect on other popular musics too, so perhaps 'nostalgia' is an afterthought or post-scriptum on a rhizome with a very real history, birth, and raison d'être. Furthermore, the techniques Hip Hop adopted—for a while unilaterally—have by now been inherited by mainstream pop producers, so Reynolds's nostalgic generalisation may be suffering from a misunderstanding of the phenomenon in its wider cross-genre implications.

Does a consideration of nostalgia, then, have any significance for the current rap practitioner? It could be argued that facing it critically brings to the forefront the *past-present* binary inherent in the sample-based aesthetic. Additionally, understanding the problematic may be helpful in drawing a line with nostalgia (i.e. the past), isolating it as a variable, so that the process of creating new sample content to serve future hip-hop development can focus on further factors.

HIP-HOP = (improvised performance + digital sampling technology) x (pre-recorded musicians + time)

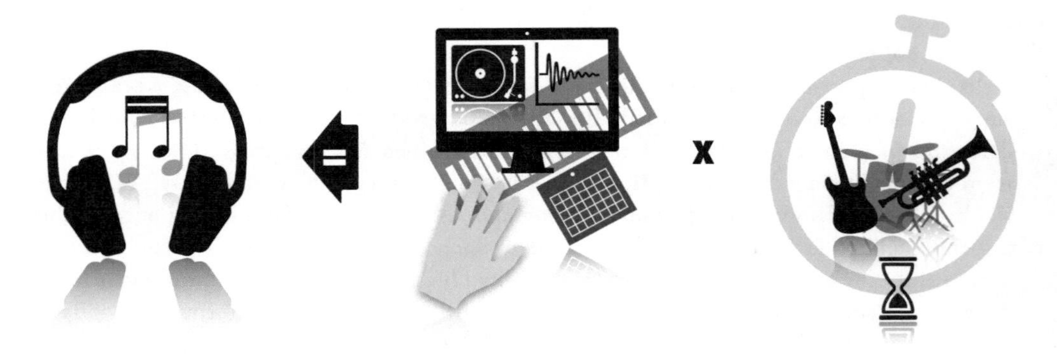

FIGURE 3.1 A schematic representation of the 'hip-hop time machine' equation.

Aesthetic problem #2: How much historicity is needed?

This brings about the question of how much historicity needs to be 'embedded' within a sample for the hip-hop aesthetic to function. It is a question that can drive a retrospective investigation of sample-based content, but also one that can inform future (re)construction. As such, it becomes theoretically important, and practically essential, should future sample-based Hip Hop continue to utilise newly constructed source content. Consequently also, defining the necessity and degree of source-content historicity will help inform the practice in a more scholastic fashion.

But what are the facets of sonic historicity that can be observed in a sample-based context? The case studies discussed above highlight a number of sonic/musical examples that help define the manifestations of this historicity in a systematic manner. There have been detailed previous attempts to provide sampling typologies (Sewell, 2013; Ratcliffe, 2014), but the focus here is somewhat different. On the one hand, the purpose of the investigation is to inform future practice, so the onus is on observing traits that are reproducible; on the other hand, this is not an attempt to account for every type of sample-use, but to do so from the perspective of what qualities infuse 'historicity' in a sample.

In the first instance, sample duration becomes an important parameter in this exploration. The longer a phonographic sample is, the more motivic information it contains. It could be argued that, conversely, a short sample—a single 'shot', 'hit', or 'stab'—focuses our attention on the sonic, granular, or layered phonographic instance.[2] Philosophically, this binary allows for a theoretical delineation between the sample as sonic instance and the sample as obvious musical or phonographic 'citation'.

Sample-based music producers may be able to 'chop' individual instrumental sounds from records (should they appear in isolation in the mono or stereo master), or alternatively opt for layered instances (such as, for example, a momentary combination of kick drum, bass note, harmonic chord, and horn stab). Access to the original multitrack data of previously released recordings has become more commonplace recently (with artists openly inviting remixers to interact with their content) and there are a number of hip-hop producers that source their samples in this fashion (as in the testimony of Amerigo Gazaway in Chapter 1). In all of these cases, the sample contains sonic information that—to the trained ear, crate-digger, or avid hip-hop fan—may point to: specific sources (single or layered instrumentation); real or artificial ambience captured during recording or applied in post-production; as well as unique sonic artefacts resulting from the recording signal flow, media used, and mixing, mastering, and manufacturing processes. Multiple sub-variables can be associated with these top-level sonic characteristics, and Figure 3.2 provides a schematic representation of essential qualities (with variables in parentheses applicable to longer segments).

Therefore, the period that the phonographic sample was captured in becomes 'communicated' even for short excerpts, because of the type of sources, spaces, equipment, and media used; but, also, the engineering and production processes applied that were typical of particular studios, production teams, record labels, and eras. Longer samples, on the other hand, may reveal all of the above, but also contain musical, rhythmical, performed, and composed utterances, which also become audible in the new context of the subsequent hip-hop production process. These add further layers of historicity to a sample, such as stylisation expressed by the compositional and arrangement choices, but also the performing idioms and musicianship carrying additional era signifiers.

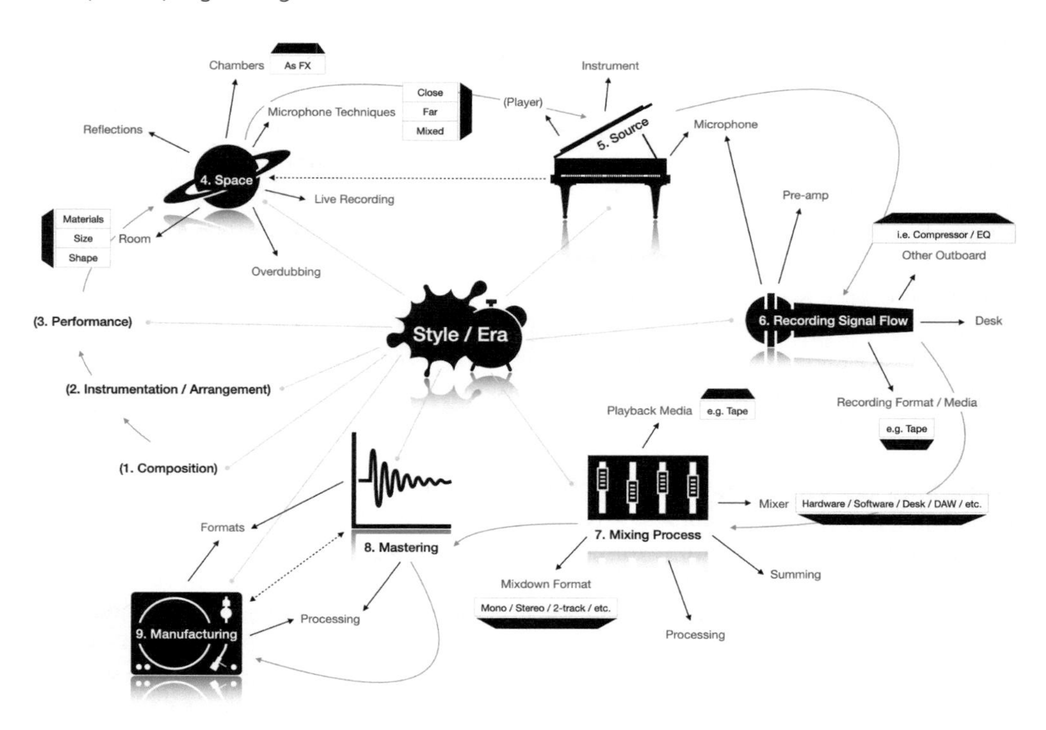

FIGURE 3.2 A schematic representation of essential phonographic sonic characteristics and related variables.

How are these observations useful to the practitioner creating new sample content that is meant to serve a sample-based aesthetic? Before even tackling the practical implications, the very process of purposely infusing historicity into—new—samples has to be analysed. Undoubtedly, there is an inherent irony in this proposition but, at the same time, from Frank Dukes to De La Soul, the notion *is* indeed practised, which necessitates a theoretical investigation.

Aesthetic problem #3: On phonographic 'magic'

The process of digital sampling can, of course, be applied to any recording, old or new, phonographic or directly recorded from an instrumental source into a digital sampler. Tellef Kvifte (2007) provides four definitions for sampling as it has been used in literature: from analogue-to-digital conversion; to the emulation of instruments by samplers; to the 'citation' of an earlier recording within a new composition; and, finally, to corrective splicing and pasting of recorded segments on analogue tape or digital formats. For a hip-hop producer, the third definition best describes the sample-based approach, particularly because a phonographic sample carries more meaning than simply a digitised acoustic vibration (and the rationale behind the process is predominantly creative rather than emulative or corrective). The distinction, though, is useful in delineating differences in workflow, as a hip-hop production built around a phonographic sample follows a very different creative trajectory to that of a production built independently of a source sample (with live instrumentation later overdubbed on top). So, a new problem that arises is that of how live instrumentation interacts with sampling.

Newly-constructed sampling content can range from recordings produced to facilitate a particular project, to ready-made content provided by sample libraries for a multitude of potential applications. There are a plethora of sample-library companies that provide wide-ranging content, from drum loops suitable for different subgenres, to live instrumentation that may fit particular styles, often accurately replicating vintage sonic characteristics mapped to very specific eras, studios, and labels. Furthermore, today's DAWs come pre-packaged with an abundance of neatly catalogued single-shot, looped, or motivic (phrase) samples and, as such, software manufacturers at least partly assume a sample-library function. Although libraries are not explicitly disregarded by hip-hop practitioners, Schloss's (2014) work observes that sample-based producers demonstrate a preference for phonographic content, showing lesser interest in ready-made content solutions.

Practitioners may be partly adhering to subcultural codes of sampling, but there are other pragmatic and aesthetic considerations to take into account. For this part of the investigation, an autoethnographic lens sheds further light onto these considerations. The methodology here consists of both composing, performing, and engineering source content to be subsequently used in a sample-based process, but also researching the historical spaces, tools, and practices behind the source references (phonographic records) pursued. In these practice-based pursuits, I have found the indefinable 'magic' of phonographic samples challenging to recreate with new recordings. As part of the historical research conducted, I have visited a number of classic studios related to the eras and records that have previously attracted me as a sample-based music producer (Chess Records in Chicago, Stax and Sun in Memphis, J&M studio in New Orleans, RCA B and Columbia in Nashville), attempting to ascertain the conditions of this 'magic': noting the spaces, microphones, signal flows, media, and equipment used, but also deciphering clues about the techniques, recording approaches, and production philosophies practised by the teams behind the recordings.

At Columbia in Nashville, staff relayed to me how Toontrack—a well-known sample-library company—utilised the facility to recreate authentic country samples for their *Traditional Country EZX* release (2016). This highlighted the oxymoron with great clarity: if there are specialists ensuring all sonic variables are adhered to in the creation of legally usable sample content, then why do hip-hop producers still opt for phonographic sources, despite the inherent copyright complications. Is the (historicity and) phonographic 'magic' more than the sum of perfectly recreated sonic—and musical—parts? Practitioners and analysts struggle to define the missing link, attributing it to a certain 'je ne sais quoi'. Citing Bill Stephney of S.O.U.L. Records, Rose (1994, p. 40) exemplifies this phonographic lure in producers' sampling rationale:

> [Rap producers have] tried recording with live drums. But you really can't replicate those sounds. Maybe it's the way engineers mike, maybe it's the lack of baffles in the room. Who knows? But that's why these kids have to go back to the old records.

Zak (2001, pp. 82–83) provides another telling account of how the phonographic process can result in such unique sonic 'ephemera', responsible for drawing the sample-based music producer in:

> The guitar's sound was bleeding into other instruments' microphones, but it had no focused presence of its own. Spector, however, insisted: this was to be the sound of "Zip-A-Dee-Doo-Dah". For it was at this moment that the complex of relationships among all the layers and aspects of the sonic texture came together to bring the desired image into focus.

This demonstrates contextual 'happy accidents', which are difficult to imagine outside of an actual record-making engagement. If there is a philosophical lesson to be acknowledged here that can then inform (re)construction, it is that of phonographic *context*. This is where sample libraries and new recordings may fall short in facilitating the sample-based artform, process, and aesthetic.

In my practice-based experiments inspired by this realisation, there was a slight shift of focus from the blues-style jamming described in Chapter 1, towards the crafting of more complete songs and structures created in pursuit of phonographic context. The following journal entry—written after an americana/folk production phase that followed the Nashville trip—exemplifies:

> There are discernible benefits to becoming fully immersed in a style/form/era of music. Listening to it almost exclusively for a period of time, picking up the relevant instruments, and engaging in record-making pursuing the auralities and the recording processes each music form requires does result in pretty convincing results—little *sound worlds* ... And I am aware I have started using the term 'songs' for some of the samples. As some of the older multitracks are sounding rather complete and multi-layered, it is time to consider the production of vocal elements for some of the samples. In order for vocals to be meaningful, though, I require concepts (rather than just 'yeahs' and 'oohs') and this has led me to some more complete songwriting recently. It is at this point I need to balance how far I go with the completion (production) of these songs, as this is not the end objective (and it is time consuming). I am consciously reminding myself that I can stop when I think I have some convincing sections. As ever, one way to establish whether the content is working in context, is to create some context... (original emphasis).

The idea was pursued within the americana/folk stylistic frame and beyond, delivering an expansive spectrum of source material. This ranged from the construction of instrumental A-B structures, through to the crafting of complete (solo) songs and demos with lyrics, songwriting collaboration with co-writers in duets and band settings, as well as progressive/fusion instrumental productions featuring more complex arrangements. Figure 3.3 provides more information on the respective source outputs and their correlation to end beats created.

But phonographic context can be influenced not only by establishing a songwriting/compositional frame, but also through the pursuit of conceptual objectives—interacting with stylistic considerations—in the sonic domain. Taken from the same americana phase, and following recording experiments combining close/mono with farther/stereo microphones upon the same acoustic sources, the journal vignette below illustrates:

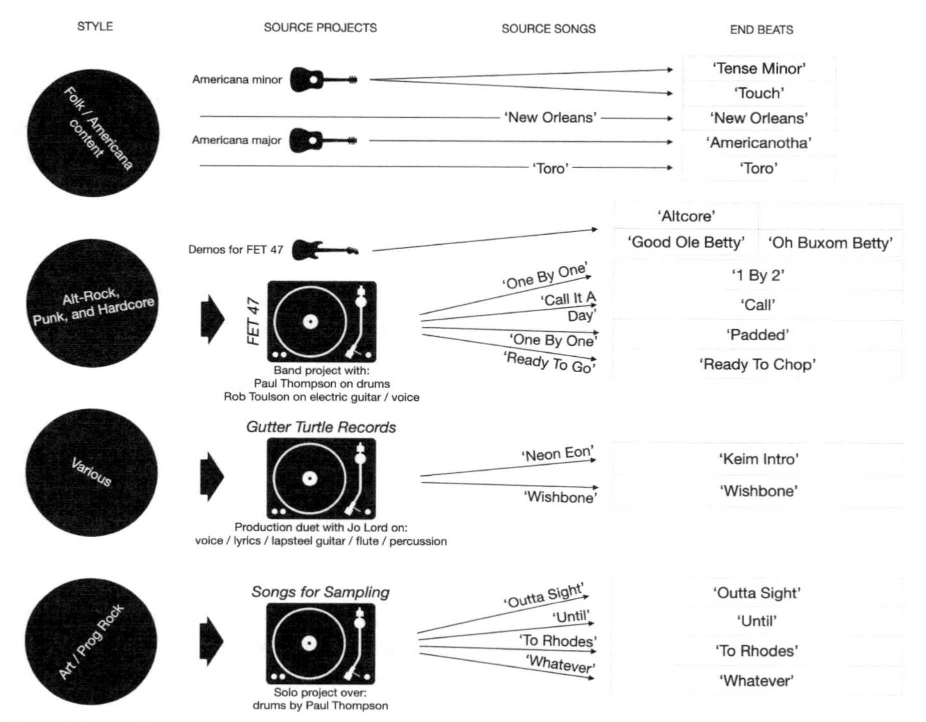

FIGURE 3.3 A schematic representation of source outputs correlated to end beats.

As a result of my previous experiments, I felt that the slightly farther, stereo perspective would provide a consistent layer of 'space' to each recording, adding to their combined sense of spatial 'glue'. I wanted the individual recordings to feel as if they had been recorded by a bunch of musicians in the same recording space.

The spatial side was only part of this pre-meditated vision towards sonic consistency. I felt it was equally important to infuse character onto the recordings by means of a 'colourful' signal flow [which consisted of a hardware tube preamplifier with an embedded compressor and equaliser on the mono source, emulations of the same preamp on both sides of the stereo microphone, and tape emulations upon all incoming channels] … I layered the mandolin first, then let it cycle, trying the ukulele on top … I listened to how the ukulele was sounding over the mandolin, going through the same preamp and EQ settings … After a practice 'lap' I tracked it as well, quickly following it up with the Martin acoustic guitar. Following the same audition, review, and recording workflow, I tracked the acoustic guitar and banjo, with minor alterations to the mic angles and preamp settings.

It is at this point that the resulting sonic 'glue' became apparent in an obvious, pleasing 'phonographic' sense: the layers of instruments with the complimentary EQ strategy, captured space, and same preamp/tape colourations started to add up to *more than the sum of the parts.* By the time I reviewed the fourth layer (the banjo)—each captured on three channels (one for the dynamic mic and two for the stereo recording)—I was listening to twelve tracks with shared characteristics and an augmented sense of shared space. There was 'glue' in their image and depth, but also in their timbral footprint, leading to a sonic sensation that felt both vintage and 'complete'; reminiscent of an era, style, location but also characterful enough that it would stick out as a potential *sample.* One particular aspect that illuminated this upon playback was how being incrementally layered, the 'wall of sound' reached a point where it blended with the vinyl crackle I had sampled and programmed as part of the MPC beat. Much less than a self-fulfilling prophecy, the instrumental mix and crackle merged into a single aural sensation, feeling like the whole thing had been lifted off an americana vinyl record.

Once I (mentally) noted this epiphany, I wanted to push it further. So, I ran the mix through (Massive Passive) mix-bus EQ, (API) compressor, and (Ampex) master tape emulations, paying tribute to the relevant sonic references with my choice of tools and settings (generally, inspired by analogue, Nashville sonic signatures). I added some Telecaster [guitar] licks using the incremental signal chtain I had finalised in yesterday's recording sessions (I recorded two tracks, one blues-funk and one psychedelic/southern-soul, focusing on developing a strong guitar signature through a combination of hardware pedals and software emulations of amp, preamp, tape, and spring reverb).[3] Finally, I tracked the ... bass ... and a Nord Wurlitzer patch through the hardware [preamp] with minor gain-staging and EQ adjustments for the same complimentary consistency, and tweaked the stereo placement of these final additions aiming for a retro "W" image.[4] As the tracks looped around, I consciously appreciated the 'wall-of-sound' effect, striking a pleasing balance between overall mix 'glue' and individual instrumental definition, albeit with shared, complementary characteristics. I asked myself the central question: would I sample this? Oh, yes, I would.

Aesthetic problem #4: The irony of reconstruction / a metamodern 'structure-of-feeling'

Phonographic context therefore appears as an essential condition in rendering samples useful to the (sample-based) hip-hop aesthetic. It could be argued that Hip Hop is borne out of the interaction between sampling processes and past phonographic content (see Figure 3.4).

But is the past—manifested as nostalgia and historicity—essential in this 'equation'? Or could convincing phonographic context—i.e. a newly constructed *record*—suffice as useable content? Some contemporary Hip Hop has indeed started quoting from more recent phonography, but the lion's share of sample-based releases focus on a more distant past. This is no surprise, as the lifespan of the style has had such a disproportionately long dependence on the sonic past that it continues to project this (past-present) temporal juxtaposition upon the majority of its outputs, almost as stylistic dogma. As Hip Hop evolves, the past may become less essential as an aesthetic qualifier, and phonographic context may become prioritised as the driver behind (the creation of) suitable sampling content. But for now, it appears that most practitioners resort to stylisation and sonics referential to past eras, in order to infuse their raw sonic materials with substantial potential for forthcoming sample-based processes.

The obvious irony observed here is that this reconstructive proposition sees the practitioner pursuing new musical content—which is (mechanically) copyright-free—whilst

FIGURE 3.4 A schematic representation of the interaction between sampling processes and (past) phonographic sonic signatures, defining the hip-hop aesthetic.

artificially infusing it with vintage sonic characteristics. This is both forward-thinking and pragmatic, but also nostalgic and 'retro-manic'. Hip-hop producers who practise this conscious duality are sonically oscillating between analogue nostalgia and digital futurism. As such, they demonstrate an awareness of Hip Hop's addiction to the phonographic past—they adhere to its nostalgic romanticism and honour this naivety—while both constructing and re-constructing: constructing new music, but *re*-constructing vintage sonic signatures. This is both naive and cynical; it puts faith in future development whilst paying homage to the dogma of historicity; and it simultaneously represents multiple dualities in a mixed, juxtaposed, and synthesised fashion, consisting of all of these polarities at once. Vermeulen and Van Den Akker (2010, p. 56) have described such a "discourse, oscillating between a modern enthusiasm and a postmodern irony, [as] metamodernism". And while the retrospective necessity in the aesthetic condition of sample-based Hip Hop is to a certain degree explained by its own historic development, technical processes, and phonographic dependences, Vermeulen and Van Den Akker's observation of a new *structure of feeling* across architecture, art, and film, (even politics) puts this (re)constructive proposition within a wider, contemporary multi-arts context.

Consequently, the notion does not simply provide a legitimisation for the conscious practice of 'irony' in this context; it rather appreciates the process as an artistic invention—or creative solution—borne out of pragmatic necessity, as *part* of an interdisciplinary movement, universal condition, or structure of feeling. It could therefore be argued that, if sample-based Hip Hop was postmodern, *reconstructive* sample-based Hip Hop is metamodern. The approaches discussed in the case studies above embrace further manifestations of metamodernism, such as: exercising multiple practitioner 'personalities' as part of the process (composer and engineer, performer of past styles, and contemporary remixer); expressing romantic compositional freedom within an Afrological, cyclic sensibility; synthesising technical precision with—and towards—the poetics of an envisioned sonic; collapsing 'time' through the juxtaposition of multiple sonic epochs; removing the historical 'distance' afforded by samples, and creating cross-genre work that offers synchronous opportunities for inter-stylistic morphing.

A practice-based anecdote

Most of the praxis, thus, in this project expresses these manifestations in the trajectory of conceptualising the dual content, and in embracing multiple roles for its construction—a process of construction that pursues a spectrum of sonic signatures (record production variables) signifying phonographic 'time' and space (eras/styles/labels/studios/locations) to invoke 'magic' or a sense of context in the source material. Armed with this conscious duality of—informed—oscillation in later parts of the journey, I began to pursue spontaneous creative events rich in contextual meaning, whilst holding an expectation of forthcoming developmental phases. The following anecdote provides a telling illustration:

A small social gathering between my partner (Jo), one of her close musical friends (Andy), and I—involving drinking and political discussion—led to a notable expressive event. The friend's commentary gradually evolved into a form of political slam/punk poetry, inspiring me to grab an electric bass and improvise a series of punk/garage-inspired lines. Andy responded with a more rhythmical and dynamic flow of delivery, we triggered a virtual drummer [in a DAW], and the spontaneous interaction evolved into a musical jam. My partner—the co-writer/singer of other source material on this project, and a spatial sound engineer—suggested that we should capture the 'energy' in the room, placing an omni-directional microphone near her listening position, and record the blend of voice and amplified bass as it emanated in the room. Anticipating that multitracking control would be beneficial in later creative stages, I handed Andy my (handheld) bullet mic, which was pre-connected to a spring reverb pedal and a fuzz stomp box (from previous, blues experiments); we put on headphones to monitor the drums (keeping us in groove), and I also miked up the bass amplifier, so that we were capturing close recordings of the (effected) voice and bass, as well as the ambient recording of the acoustic performances in the room. In haste, I chose virtual [console] pre-amplifiers on the software mixer bridging the real sonic world with the computer conversion (the recording interface) that matched the aggressive energy of the performances (emulating Neve consoles). We recorded constantly for two cycles of the virtual drum arrangement (approximately 18 minutes in total).

[A few days after this event] I revisited the recordings and layered more instruments… Inspired by the punk/garage attitude of the instantaneous performance captured (it reminded me of the energy in early Velvet Underground records), I recorded layers of overdriven electric guitars and upright piano stabs (all in the same room, pursuing a post/retrospective spatial blend with the previous performances). Not having worked with a real drummer in the initial jam, I felt that the only unrealistic timbre in the resulting mix was that of the virtual drums. I connected the computer to the Bluetooth speaker positioned on top of the piano, singled out the drum mix, played it out loud, and recorded it from some distance with the same omni mic deployed on day one. I ran the mic signal through my vintage hardware compressor/pre-amp, 'squashed' it hard, and boosted the extremes of the frequency spectrum. It began to sound like a room microphone tracking a loud drum performance in a drum booth. The song/jam was mixed deploying software approximations of technologies characteristic of the era (multitrack and master tape, console, and effect emulations).

The vignette above provides a set of phonographic 'sparks' resulting in contextual 'ephemera', as well as an illustration of both naïve (in flow) and calculative (analytic) actions characteristic of the metamodern oscillation.[5] The 18-minute 'record' has been utilised in four different beats and at varying degrees in the end project. Sampled and chopped as an intact stereo master, it has provided the foundational samples for 'Boxing'. Elements of the isolated voice—nevertheless effected for the context of the source multitrack (and additionally processed in the context of the mix/master process)—have been used in three further beats: 'Keim Outro', '1960s 2', and 'Genitive', which will reappear in discussions concerning the remainder of their foundational samples in later chapters. 'Boxing' showcases the source material in a more exposed fashion at 2:30–3:00 (and again at 4:29 and until the very end of the sample-based production).

Conclusions

The chapter's narrative commenced with a hypothesis that the sample-based hip-hop aesthetic is borne out of the interaction of sampling processes with past phonographic signatures, before questioning the nature and degree of the manifestation of the past as a variable in this creative equation. Acknowledging a number of contemporary approaches where hip-hop practitioners create new content in order to facilitate sample-based processes, the investigation has consequently examined the variables that enable an effective interaction between newly created content and beat-making. The aim has been to theorise on this dynamic, arming future practitioners with a better understanding of the aesthetic implications in dealing with both phonographic and newly created sampling content—so that referential sonic 'objects' (Moylan, 2014) can be (re)constructed, aiding the future evolution of the (sub)genre.

Looking at representative practitioners deploying a number of alternative contemporary approaches as a way to innovate and negotiate the pool of available sampling material, I have theorised on four areas of aesthetic concern: the function of sonic/musical nostalgia; the infusion of 'historicity' onto source content; the notion of phonographic 'magic'; and the identification of this reconstructive proposition as metamodern practice. In comparing phonographic samples to newly recorded source material or sample-library content, a number of differences have become apparent, which point to the techno-artistic processes deployed in the construction of vintage sonic material. These, nevertheless, can arguably be re-constructed to close proximity as exemplified by the meticulous reverse-engineering of both sample-library companies and practitioners alike.

Therefore, the missing link in explaining sample-based producers' preference for phonographic content, and the unquantifiable 'draw' towards it, may be situated in the ephemeral manifestations of cultural resonance that result from the complex interactions and chaotic dynamic of the record-making process: sounds and utterances resulting from phonographic *context*. Perhaps the most promising potential for the future of a sample-based approach (in Hip Hop and beyond) lies in the exponential promise of 'making records within records', consciously *squaring* phonographic context, and synthesising the paradigm with that of a metamodern 'structure of feeling'. Far from simply adopting a fitting reconstructive frame, the empowerment for the practitioner in this synthesis stems from the realisation that the simultaneous irony of (postmodern) reconstruction merged with the enthusiasm of (modern) creation represents indeed a universal and interdisciplinary cultural paradigm. As a result, a subset of contemporary rap artists and producers may just be engineering new or future cultural objects, whilst consciously *and* naively entertaining their nostalgic predisposition towards stylisations resulting from the interaction of sampling processes with phonographic ephemera. This way, the oscillation between sonic pasts and hip-hop futures may result in a collapse of time and historical 'distance' via the very synthesis of vintage production techniques and sample-based processes. As Whalen (2016) identifies in the work of Frank Dukes: "By reverse-engineering the art of flipping samples, Feeney is looking at the past, present, and future simultaneously".

Recommended chapter playlist
(in order of appearance in the text)

'Tense Minor'
'Touch'
'New Orleans'
'Americanotha'
'Toro'
'Altcore'
'Good Ole Betty'
'Oh Buxom Betty'
'1 By 2'
'Call'
'Padded'
'Ready To Chop'
'Keim Intro'
'Wishbone'
'Outta Sight'
'Until'
'To Rhodes'
'Whatever'
'King's Funk'
'Psychodelic'
'Boxing'
'Keim Outro'
'1960s 2'
'Genitive'

Notes

1 The Old School period in Hip Hop is typically regarded as ranging from 1979 to 1982, signified by releases such as 'Rapper's Delight' (Sugarhill Gang, 1979) in the live arena, 'The Adventures Of Grandmaster Flash On The Wheels Of Steel' (Grandmaster Flash, 1981) representing turntablism, and 'Planet Rock' (Afrika Bambaataa & The Soul Sonic Force, 1982) and 'The Message' (Grandmaster Flash & The Furious Five, 1982) exemplifying Electro and Electro-funk, respectively (see, Exarchos, 2020, for a discussion on the function of synthesisers in Hip Hop's trajectory and eras).

2 Citing Gillespie, Williams (2010, p. 21) delineates the sample-based use of the two, respectively, as "syntagmatic" and "morphemic sequencing".

3 These respectively led to the 'King's Funk' beat, in reference to Albert and Freddie King; and the 'Psychodelic' beat, referencing work such as *The Baby Huey Story – The Living Legend* (Baby Huey, 1971).

4 Roey Izhaki (2008, p. 71) defines a W image as a common "stereo spread imbalance" that can be found in "a mix that has most of its elements panned hard-left, center and hard-right". He offers the verses of OutKast's 'Hey Ya!', from *Speakerboxxx/The Love Below* (2003), as a telling, retro-inspired, example.

5 Keith Sawyer (2017, p. 96) asserts that "creative sparks are always embedded in a collaborative process, with five basic stages" (preparation, time off, the spark, selection, and elaboration); and even when sparks appear as individual insights, they are "deeply embedded in the knowledge and social interactions of the preparation and time-off phases, and [they build] on sparks that others have had". Collaboration in Sawyer's terms thus includes thinking and acting as part of an invisible network, which he refers to as "the collaborative web" (Sawyer, 2017, pp. 207–233). Citing art sociologist Howard Becker (1982, p. 25), Vera John-Steiner (2006, p. 4) agrees that even in supposedly 'lone' artforms such as painting and poetry "the artist ... works in the centre of a network of cooperating people whose work is essential to the final outcome". This echoes Zak's idea of the aforementioned phonographic "resonant frame", in which record collections "represent historical documents and *instruments of instruction*" (my emphasis) for record producers.

Bibliodiscography

44th Anniversary of the Birth of Hip Hop (2017) *Google*. Available at: https://www.google.com/doodles/44th-anniversary-of-the-birth-of-hip-hop (Accessed: 11 August 2017).

Afrika Bambaataa & The Soul Sonic Force (1982) *Planet Rock* [Vinyl, 12"]. US: Tommy Boy.

Baby Huey (1971) *The Baby Huey Story – the Living Legend* [Vinyl LP]. US: Buddah Records.

Beastie Boys (1989) *Paul's Boutique* [CD, Album]. US: Capitol Records, Beastie Boys Records.

Becker, H.S. (1982) *Art worlds*. Berkeley: University of California Press.

Beer, D. and Sandywell, B. (2005) 'Stylistic morphing: Notes on the digitisation of contemporary music culture', *Convergence: The International Journal of Research into New Media Technologies*, 11(4), pp. 106–121.

Chang, J. (2007) *Can't stop won't stop: A history of the hip-hop generation*. Reading, PA: St. Martin's Press.

Exarchos, M. (2020) 'Synth sonics as stylistic signifiers in sample-based Hip-Hop: Synthetic aesthetics from "Old-School" to Trap', in N. Wilson (ed.) *Interpreting the synthesizer: Meaning through sonics*. Newcastle upon Tyne: Cambridge Scholars Publishing, pp. 36–69.

Grandmaster Flash (1981) *The Adventures of Grandmaster Flash on the Wheels of Steel* [Vinyl LP]. US: Sugar Hill Records.

Grandmaster Flash & The Furious Five (1982) *The Message* [Vinyl LP]. US: Sugar Hill Records.

Izhaki, R. (2008) *Mixing audio: Concepts, practices and tools*. Oxford: Focal Press.

John-Steiner, V. (2006) *Creative collaboration*. Oxford: Oxford University Press.

Kvifte, T. (2007) 'Digital sampling and analogue aesthetics', in A. Melberg (ed.) *Aesthetics at Work*. Oslo: Unipub, pp. 105–128.

Law, C. (2016) *Behind the Beat: J.U.S.T.I.C.E. League, HotNewHipHop*. Available at: https://www.hotnewhiphop.com/behind-the-beat-justice-league-news.23006.html (Accessed: 6 September 2017).

LeRoy, D. (2006) *Paul's boutique*. New York: Continuum (33 1/3).

Marshall, W. (2006) 'Giving up hip-hop's firstborn: A quest for the real after the death of sampling', *Callaloo*, 29(3), pp. 868–892.

Morey, J.E. and McIntrye, P. (2014) 'The creative studio practice of contemporary dance music sampling composers', *Dancecult: Journal of Electronic Dance Music Culture*, 1(6), pp. 41–60.

Moylan, W. (2014) *Understanding and crafting the mix: The art of recording*. 3rd edn. Oxon: CRC Press.

OutKast (2003) *Speakerboxxx/the Love Below* [2xCD, Album]. Europe: Arista.

Public Enemy (1988) *It Takes a Nation of Millions to Hold Us Back* [CD, Album]. UK & Europe: Def Jam Recordings.

Public Enemy (1990) *Fear of a Black Planet* [CD, Album]. Europe: Def Jam Recordings.

Ratcliffe, R. (2014) 'A proposed typology of sampled material within electronic dance music', *Dancecult: Journal of Electronic Dance Music Culture*, 6(1), pp. 97–122.

Reynolds, S. (2011) *Retromania: Pop culture's addiction to its own past*. New York: Faber and Faber.

Rose, T. (1994) *Black noise: Rap music and black culture in contemporary America*. Hanover, NH: University Press of New England (Music/Culture).

Sawyer, K. (2017) *Group genius: The creative power of collaboration*. Revised and updated. New York: Basic Books.

Schloss, J.G. (2014) *Making beats: The art of sample-based Hip-Hop*. Middletown, CT: Wesleyan University Press (Music/Culture).

Sewell, A. (2013) *A typology of sampling in hip-hop*. Unpublished PhD thesis. Indiana University.

Sugarhill Gang (1979) *Rapper's Delight* [Vinyl LP]. US: Sugar Hill Records.

The Roots (1993) *Organix* [CD, Album]. US: Remedy Recordings, Cargo Records.

Thompson, A. and Greenman, B. (2013) *Mo' meta blues: The world according to Questlove*. 1st edn. New York: Grand Central Publishing.

Toontrack Traditional Country EZX released (2016) *rekkerd.org*. Available at: https://rekkerd.org/toontrack-releases-traditional-country-ezx/ (Accessed: 2 October 2017).

Van Poecke, N. (2014) 'Beyond Postmodern Narcolepsy', *Notes on Metamodernism*, 4 June. Available at: http://www.metamodernism.com/2014/06/04/beyond-postmodern-narcolepsy/ (Accessed: 5 September 2015).

Vermeulen, T. and Van Den Akker, R. (2010) 'Notes on metamodernism', *Journal of Aesthetics & Culture*, 2(1), pp. 56–77.

Weingarten, C.R. (2010) *It takes a nation of millions to hold us back*. New York: Continuum (33 1/3).

West, K. (2016) *The Life Of Pablo* [Digital Release, Album]. G.O.O.D. MUSIC, Def Jam Recordings.

Whalen, E. (2016) *Frank Dukes Is Low-Key Producing Everyone Right Now, The FADER*. Available at: https://www.thefader.com/2016/02/04/frank-dukes-producer-interview (Accessed: 6 September 2017).

Williams, J.A. (2010) *Musical borrowing in hip-hop music: Theoretical frameworks and case studies*. Unpublished PhD thesis. University of Nottingham.

Zak III, A.J. (2001) *The poetics of rock: Cutting tracks, making records*. Berkeley: University of California Press.

PART 3

(The magic of sample-based) Production

4

PHONOGRAPHIC GHOSTS AND META-ILLUSIONS IN CONTEMPORARY BEAT-MAKING

From '(al)chemists' and 'wizards' of the beat to the 'magic' of phonographic sampling in hip-hop music, the practice and literature surrounding sample-based music production are inundated with supernatural references. But as we have seen in Chapter 3, sample-based music creation has also been criticised as "a mixture of time-travel and séance" (Reynolds 2011, p. 313), where subjects from the past are unwillingly manipulated by contemporary music makers. Whether 'magical' vocabulary is mobilised in these contexts in a complimentary or critical sense, it is important to question why it is used to describe musical phenomena and, specifically for the focus of this project, how it applies to sample-based music creation.

At first glance, it is easy to see how the amazement resulting from musical feats or pleasing aesthetic results (in any artform) can lead to exclamations of awe and an elevation of the artist's skill to supernatural dimensions. It is important, however, to investigate more explicitly the conditions under which a musician—and for the purposes of this chapter, a beat-maker—becomes a magician in the eyes (ears) of their audience, as well as the implications of this transformation for both artistic effect and the audience experience. Furthermore, it will be useful to explore whether the frequent use of magical or supernatural characterisations simply substitutes complimentary (or critical) language directed at artists, or whether there is something more profound about their unanimous and universal usage. This will also provide the opportunity to explore whether such terms have become pivot mechanisms in popular music parlance diverting attention away from the serious study of artistic phenomena, when there are unexplained aesthetic effects taking place that warrant more careful examination.

Conditions and parallels

A logical position for the investigation to start from is the pursuit of the conditions necessary for magic in a sample-based musical context to occur. But first 'magic' itself requires a definition relevant to an artistic context. The focus here will remain on one understanding of

DOI: 10.4324/9781003027430-9

magic as 'stage' or 'performance' magic, examining the conditions necessary for performance magic to occur, before drawing parallels to sample-based music creation. In doing so, I will demonstrate that music and magic work as reciprocal metaphors not only because music is frequently compared to magic, but also because stage magicians consistently use time-based, musical metaphors when explaining their practice. Furthermore, the obstacle of comparing a predominantly performable artform (stage magic) with a mediated one (sample-based music production) will be dealt with, at large, through a discussion of performable utterances that can be identified on the latter as an expression of traditional turntable practices.[1]

Starting from a more generic notion of magic, Oxford Dictionaries (2018) define it as "[t]he power of apparently influencing events by using mysterious or supernatural forces". Yet, from Houdini to Penn and Teller, performance magicians have dedicated much of their lives' efforts to exposing fraudulent claims towards the supernatural, and educating their audiences about the skill and effort required in delivering effective performance magic. Lamont and Wiseman (1999, p. xvi) claim that "magic, properly performed, is a complex and skillful art", while Vance (1985, cited in Wilcock, 2015, p. 40) describes magic as "a practical science, or more properly, a craft". Penn and Teller (cited in Zompetti and Miller, 2015, p. 11) go as far as to expose their methods on television because, according to them, "illusions are just illusions"; and Fitzkee (1945/2009, cited in Zompetti and Miller, 2015, p. 8) agrees that "what makes a magic trick great... is *performance*" (original emphasis). Claims such as these are echoed throughout the world of performance magic, demonstrating that a reading of terms and ideas referring to magic through the lens of craftsmanship, mastery, and skill (rather than an acceptance of the supernatural) may render parallels that are more useful for the study of what is referred to as 'magical' in other performing arts.

Teller (cited in Leddington, 2016, p. 256) describes magic as "a very, very odd [art]form" and Leddington (2016, p. 254) agrees that "magic does not fit neatly into our usual aesthetic categories". This could also be said about the aesthetics of both record production more generally and sample-based record production more specifically, since the latter is borne out of the layering, manipulation, and juxtaposition of previously made phonographic constructs. In relation to hip-hop music, Schloss (2014, pp. 72–78) explains that "the idea of sampling as an aesthetic ideal may appear jarring to individuals trained in other musical traditions, but it absolutely exemplifies the approach of most hip-hop producers", adding that "this preference is not for the act of sampling, but for the sound of sampling: It is a matter of aesthetics".

In his critique of sample-based music as "séance fiction", "the musical art of ghost co-ordination and ghost arrangement", a process that "doubles [recording's] inherent supernaturalism", and the resulting "collage [as] a musical event that never happened", Reynolds (2011, pp. 312–314) described the aesthetic conundrum quite acutely. Reynolds's critical stance towards sample-based music allows him a distanced analysis of the aesthetic phenomena on hand and, although rap practitioners may not share his disdain, he does, however, offer some helpful analogies between the two artforms: specifically, the manipulation of others' energies, the condition of distance, and the effect of unwillingness (on the side of subjects or audiences).

Lamont and Wiseman (1999)—assuming both practitioner and scholarly roles—provide a systematic account of the conditions necessary for effective performance magic, allowing us to draw more insightful parallels between the mechanics of the two artforms. Their findings appear surprisingly apt at describing the mechanics of music (both its performance and creation), especially if read with a focus on the interdependence between a performer's *method* and its *effect* on an audience (the listener). In their book, *Magic in theory*, they claim that "[s]uccess [in the performance of magic] requires that the spectator experience[s] the effect while being unaware of the method" (Lamont and Wiseman, 1999, p. 29), a dynamic that Zompetti and Miller (2015, p. 8) echo as the precedence of "the wonder of the occurrence" over the "mechanics of a trick". Indeed, Leddington (2016, p. 258) attests that "the magician has to... 'cancel' all the methods that might reasonably occur to you" and Hay (1972, p. 2, cited in Zompetti and Miller, 2015, p. 12) sums up that the "secret of conjuring is a manipulation of interest". What's noteworthy in *Magic in theory*, however, is Lamont and Wiseman's frequent discussion of rhythm and timing as crucial devices in interest manipulation, which resonate sympathetically with sample-based Hip Hop's preoccupation with rhythm and groove. The authors describe: novelty, "sudden sound(s)", "change of pace", (relative) movement, contrast (Lamont and Wiseman, 1999, pp. 40–41); as well as highlighting the "moment of effect" over "the moment of method", and using rhythm, punctuation, and (on-)beat and "off-beat" moments as essential tools in physical misdirection and time-based attention control (Lamont and Wiseman, 1999, pp. 46–53).

The cyclic structure of sample-based Hip Hop similarly depends on interest manipulation through rhythmic and textural dynamics: cuts and stops of the beat, dynamic manipulation of found samples, use of sound effects, and spatial effect processing. If we take Gang Starr's 'Code of the Streets' (from *Hard to Earn*, 1994) as a classic example, producer DJ Premier juxtaposes the exposed—and slightly sped up—introductory drum beat from 'Synthetic Substitution' (1973) by Melvin Bliss over the pitch-shifted instrumental introduction to 'Little Green Apples' (from *Extra Soul Perception*, 1968) by Monk Higgins. He then deploys Beside's *Change the Beat* (1982) as a source for his turntable scratching and manipulation that constitutes the track's chorus. Throughout the track, Premier *cuts* the instrumental sample in time with the beat at key moments to create dynamic interest (e.g. at 0:18, before rapper Guru starts his first verse), while the end phrases of his scratching on later choruses are prolonged using a delay effect, their repeats fading out into subsequent verses. The track also features extraneous amounts of vinyl noise (underlining the connection to turntablism already evident in the scratching), while the 'Synthetic Substitution' beat sounds reinforced through the use of equalisation, potentially additional drum layers, and a prominent level placement in the mix. Premier's production here illustrates how the organisation and manipulation of full phonographic layers—rather than individual instrumental elements—results in striking dynamic, textural, and rhythmic effects, which he also deploys to provide the contextual materials behind Guru taking centre stage in the verses. On a later production, such as 'Deadly Habitz' (from album *The Ownerz*, Gang Starr, 2003), his manipulation of Steve Gray's 'Beverly Hills' (from *Relax*, 1979) demonstrates how a beat-maker can create reimagined phrases out of found phonographic segments, creating rhythmical interactions between the (recorded) gestures and those included in the sampled content. As such, motion is perceptible on multiple levels while various layers can be brought to the listeners' attention though relative level balancing (mixing), timbral and spatial enhancements (equalisation, use of delay and reverb effects), or the performative manifestations of the producer's actions. The cyclical nature of sample-based Hip Hop follows Afrological priorities (Lewis, 2017), which Rose (1994, p. 83) acknowledges as a characteristic of "black cultural traditions and practices" expressed in manifestations of "openness, ruptures, breaks and forces in motion". The listener's attention is directed towards sonic, rhythmic, and lyrical invention expressed over a familiar loop; which is no different to the phenomenon of diverting spectators' foci to *effect* in performance magic, using the mechanisms of "naturalness", "consistency", "familiarisation", reinforcement, continuity, and subtlety—conditions Lamont and Wiseman (1999, pp. 60–74) identify as essential for psychological (mis)direction.

Structure, control, and subgenre

A further parallel that can be drawn between performance magic and sample-based record production relates to the structure of the artistic exposition and the use of 'raw materials' to construct it. Reinhart (2015, p. 26) cites the opening sequence of mystery thriller *The Prestige* to describe the typical "praxis of the commercial magical show":

> Every great magic trick consists of three parts or acts. The first part is called "The Pledge" in which the magician shows you something ordinary... The second act is called "The Turn". The magician takes the ordinary something and makes it do something extraordinary. Now you're looking for the secret... That's why every magic trick has a third act, the hardest part ... the part we call "The Prestige".
>
> *(from the opening sequence of The Prestige, cited in Reinhart, 2015, p. 25)*

Reinhart (2015, p. 26) explains that "in a great magic performance there are always two gaps - one between the 'Ordinary Something' and the 'Unexpected' and the second one between the 'Unknown' and a magically restored order". If we listen to a number of sample-based hip-hop tracks, such as 'Lightworks' by J Dilla (from *Donuts*, 2006), or 'Filthy (Untouched)' by Madlib (under his Beat Konducta alias, from *Vol. 1–2: Movie Scenes*, 2006), we can identify a very similar structural idea. The producers initially expose relatively unprocessed (albeit pitch-shifted and somewhat equalised) phonographic samples—segments from Raymond Scott's 'Lightworks' (from *Manhattan Research Inc.*, 2000) amongst others, and Vivien Goldman's 'Launderette' (from *Resolutionary*, 2016), respectively—letting us *in on the trick* they are about to 'perform' so to speak. This is in order to then mesmerise us with their abilities to manipulate, truncate, loop, and re-order ('chop') their phonographic sources—a process collectively known as 'flipping' in hip-hop practice. Skilful rap producers such as J Dilla and Madlib are capable of presenting reimagined sequences and sonic constructs substantially altered from their phonographic origins, therefore the introductory 'pledge' plays in their favour, demonstrating a notable 'gap' or 'turn' into the 'unexpected'. Note that the titles the producers choose for their resulting sample-based creations are consistent with their process—J Dilla keeps the name 'ordinary' (the same), while Madlib hints at the 'flipping' strategy by naming his track 'Filthy (Untouched)', rather than 'Launderette'. It could be argued that the 'restored' order comes in the form of establishing a new cyclic structure held together by the timbral and rhythmic coherency of the new main 'hook' (loop) that drives the rest of the production. The limitation of having to work with a small number of sampled phonographic instances (rather than unlimited instrumentation), necessitates the construction of narrative and the retention of interest predominantly through the manipulation of an 'ordinary something'—a raw sonic source in the context of a sample-based hip-hop structure. Although this process becomes mediated on record, it owes much to the aural tradition of turntablism, being communicated to listeners via this developmental exposition of raw materials and subsequent sonic constructs. The parallel exposes a fundamental structural insight in the art of 'flipping' samples in Hip Hop, highlighting not only a surface characteristic, but identifying instead an essential mechanism in engaging and retaining listener interest, whilst authenticating the producer as 'performer' in control.

Demonstrating control, furthermore, appears as another fundamental condition before a performer can acquire 'magical' status within performance magic or beyond. Lamont and Wiseman (1999, p. 64) suggest that "[a]uthority brings control, and control of the situation can allow the magician to set the conditions"—an axiom that also applies to beat-makers and their demonstration of 'chopmanship' or 'wizzardy' over phonographic samples. Loshin (2007, cited in Zompetti and Miller, 2015, p. 10) confirms that "[t]he metanarrative of magic is tied up with the notion of control… control of the natural world". The notion can be expanded to include scientists, as they can also discover, apply, and display methods of control over various forms of physical energy. For 18th-century physics professor and stage magician Étienne-Gaspard Robert "the magical show was simply a lesson in applied physics, performed to amaze and educate his audience" (Reinhart, 2015, p. 32). In the case of Thomas Alva Edison, his 'power' over acoustic energy, channelled via the invention of the phonograph, earned him the nickname of the "Wizard of Menlo Park": through it, he was capable of "transforming life into abstract signals and playing them back… [allowing] us to hear voices from people absent or long gone" (Reinhart, 2015, p. 27). In the case of beat-makers, too, the control over sonic materials, demonstrated through the art of flipping samples, is a form of control over musical (rhythmical, textural, motivic) relationships, but also stylistic invariables. This is exemplified by beat-makers', DJs', and mashup remixers' aliases and characterisations, such as The Alchemist, DJ Cut Chemist of Jurassic 5, and Amerigo Gazaway: 'chemist' of the mashup, as previously described in press (Caldwell, 2015). In a more general sense, Kugelberg (2007, p. 31) sees all of hip-hop musicking as an artform comparable to alchemy or magic: "With hip hop, born in the Bronx, these guys created something out of nothing. That's amazing. That's alchemy. That's magic". For Reinhart (2015, p. 31), the ability to create 'something out of nothing' also characterises scientists, inventors, and magicians—this is because they are "exponents of the same mind set [as they] have learned to deal with the phantasmatic space of the unknown in a creative way".[2] A quality that perhaps applies to all music makers, but one that appears even more fitting for sample-based producers due to the materiality (rather than abstraction) inherent in the nature of control manifested over their sonic objects.

Control in this sense is understood as an intrinsic condition of both performance magic and sample-based music production describing the actions the performer exercises over the materials or objects deployed. The implication is that the audience (or listeners) are entertained by observing (or listening to)—and sometimes interacting with—the manifestations of control (effect), as exercised by the performers (method). Robert-Houdin (1906, cited in Reinhart, 2015, p. 31) categorises Modern Magic into classes according to such intrinsic characteristics (i.e. what subjects are used and what actions are performed). For example, one of the classes he identifies, "Experiments in Natural Magic", is described as "[e]xpedients derived from the sciences and which are worked in combination with feats of dexterity, the combined result constituting conjuring tricks" (Reinhart, 2015, p. 31). The definition sounds analogous to how sample-based record production could be described as a (sub)genre: *exponential phonographic illusions derived from sonic phenomena (psychoacoustics) and the manifestation of producer (originally turntablist) dexterity over phonographic sound objects—the combined result constituting* 'supernatural sonic collages' (to echo Reynolds again).

As much as it is useful to classify artforms through the lens of practice and the materials used (i.e. intrinsically), the resulting effect(s) (appreciation, entertainment) cannot be fully comprehended without considering audience perception and the context surrounding recipients (culture and mediation). Landman (2013, p. 47) theorises on how different genres of theatrical or stage magic frame performance "on a different contract between the performer and the audience, the discourse used during performance and the effect on the audience both in terms of its perception of what has transpired and the personal meaning attached to the effect". He expands with a fitting analogy:

> Like the different strands in other performing arts (music, drama, comedy), these genres have distinctive communities and sub-cultures, as practitioners try to establish hegemony of one form of performance magic over others, or seek to construct separate identities around their stage persona and approaches to performance magic.
>
> *(Landman, 2013, p. 48)*

Similarly, the work of sample-based beat-makers, through its historical association to turntablism, is rooted in approaches to music production congruent with an evolving stylistic contract between producers and fans, requiring methods of control from the side of the makers, in order to create coherent sonic experiences for the recipients. As with any evolving stylistic contract, the successful producers manage to challenge recipient interest by balancing adherence to aesthetic criteria (genre rules), whilst innovating; but the parallel force in action is the evolution of audience perception itself. As Cohen (1994, cited in Zompetti and Miller, 2015, p. 12) reports on Penn and Teller, they are performers "willing to acknowledge...that the culture is savvy to magic"; just as musical/hip-hop culture is savvy to sample-based utterances and sampling technology. Williams (2014, p. 193) confirms: "hip-hop as a genre presupposes an un-concealed intertextuality which is part and parcel of its aesthetics. Much of this has to do with hip-hop communities' expectations (its 'generic contract')".

However, the adoption of developing mediation technologies into any performing art adds another crucial factor that negotiates how the artform is framed. The following section will widen the focus of the discussion to the totality of record production, in order to explore parallels between mediation effects upon phonography and performance magic. Sample-based record production will then be contextualised as a subgenre defined specifically by sampling technology affordances. The discussion hypothesises that if the art of record production evolved from documentarian capture *(as real)* to the construction of sonic illusions *(hyper-real)*, then sample-based Hip Hop introduces the notion of an exponential juxtaposition of multiple illusions *(meta-real)*. The phenomenon mirrors the evolution of audience perception in performance magic from: 'seeing is believing' *(magic as the occult)*; to illusionism *(suspended disbelief)*; to TV illusionism *(challenging the interaction between performance and production)*. The interaction of technology with both artforms has implications for their 'frame' in relation to liveness, authenticity, and defining aesthetic criteria, necessitating also a theorising of the 'meta-*effect*' that is possible through mediation. A schematic representation of the parallel streams of evolution for both artforms and the respective audience perception can be seen in Figure 4.1, including associated technological variables, tools used in directing audience attention, and related processes.

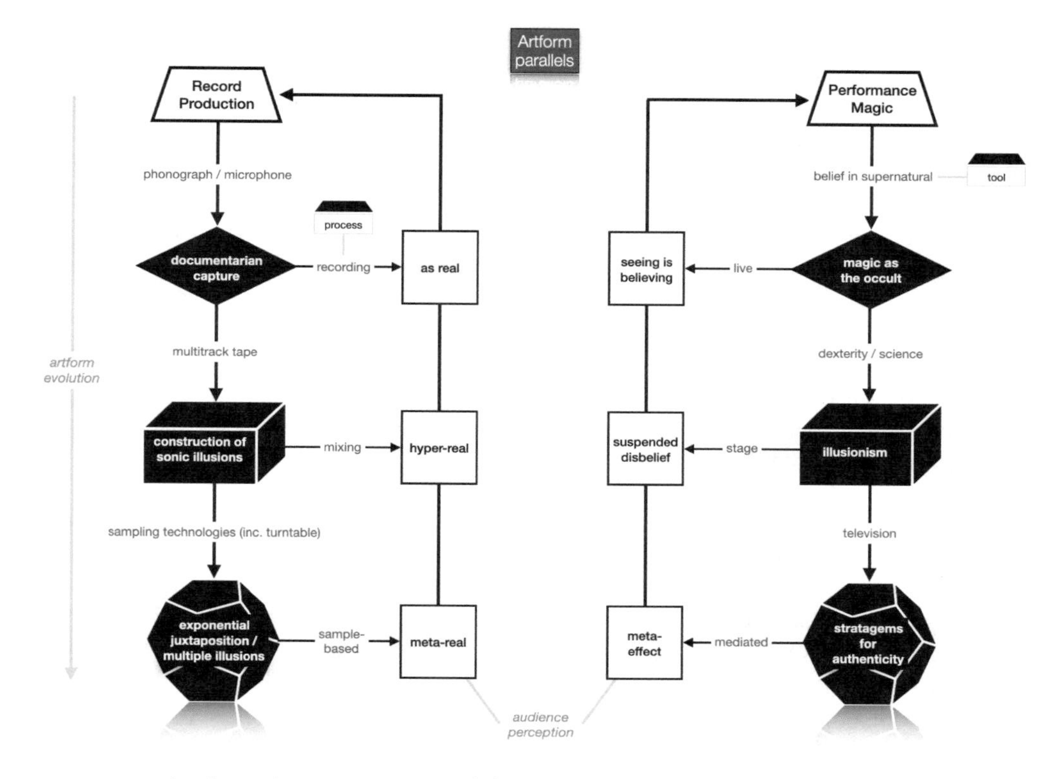

FIGURE 4.1 A schematic representation of the parallel evolutionary streams for record production and performance magic.

Technology, mediation, and the alchemy of beat-making

Landman (2013, p. 52) proposes a convincing categorisation of genres and subgenres in performance magic according to the methods (materials), effects, and frame adopted, but also the contract, engagement, and discourse that can be observed between performer(s) and the audience. In the case of the overarching categories of "magic", "mentalis", and "mystery entertainment", he finds that the former two share a—contemporary—frame relating to the production of inexplicable effects based on unknown methods, while in the latter there is an implicit claim towards the supernatural (Landman, 2013, p. 56). As we have seen above, however, the delineation away from supernatural claims characterises most *modern* magic, mirroring the growing 'savviness' of a contemporary audience. The evolved perception identifiable in audience culture is understandably the result of multiple forces;[3] but it is also a consequence of conditioning borne out of the ongoing discourse between performers and audiences. As Leddington (2017, p. 34) notes: "[t]hat the sign reads 'Magic Show' allows us to feel a measure of safety and control in the face of what might otherwise be a frightening experience: encountering an apparent violation of natural law"; and he maps audience reaction to a behavioural mechanism he describes as an "activation of the 'magic' genre script"—something Landman (2013) sees as the unspoken contract framed between performer and participants.

The obvious parallel with record production is a similar 'savviness' amongst listeners about recording technologies, creative possibilities, and resulting effects, which create a range of artistic expectations throughout the history of the artform and across different (sub)genres. Much has been written about the advent of multitracking and its effect on the aesthetics of record production. By enabling the manipulation of separate layers of instrumentation in post-production (see, for example: Katz, 2010; Horning, 2013; Jarrett, 2014) and—through that—the *staging* of balance, timbral, and spatial illusions (see: Lacasse, 2000; Zagorski-Thomas, 2010; Liu-Rosenbaum, 2012), it has transcended the pursuit of a purely documentarian approach to sonic representation. Jarrett (2014, p. 113) aptly describes:

> Around 1967, recordings changed. From that point on, they were almost never actual records of single musical events—they became instead, almost always, composites of many musical events–"virtual" records. The performances heard on records were more constructed than caught.

Nevertheless, the 'genre script' across eras and sub/genres has continued its differentiation beyond the initial affordances of multitrack tape, in response to a plethora of consumption trends and developing media. Zagorski-Thomas (2010) has demonstrated the effect that the interplay between idealised and actual consumption spaces has had on the development of mixing strategies for rock and disco (and eventually dance music), while Thomas Vendryes (2015) has highlighted the socio-economic context within which King Tubby pioneered the notion of the 'remix' (and, simultaneously, the subgenre of Dub). It could be said, however, that, collectively, the art of record production, in all of these pursuits since the invention of multitrack tape and until the dawn of sampling technologies, had been freed from the limitation of the 'real' (performance representation) and was allowed to explore and build 'hyper-real' sonic constructs. Of course, representational or documentarian outputs have continued to be produced, with particular genres placing high value on the least amount of mediation over authentic performances.[4] Furthermore, even when attempting to represent a performance with minimum mediation, a certain amount of distortion of the acoustic representations is inevitable due to the recording and post-production techniques and tools employed (Zagorski-Thomas, 2018, pp. 13–24).

In performance magic, the shift from a perception of conjuring effects 'as real' (stemming from supernatural powers) to the adoption of *suspended disbelief* as a condition enabling 'illusory' entertainment, seems to have been brought about by a philosophical set of conditions (rather than a particular set of technological affordances). Nevertheless, it is interesting to note the implications for magic once its performance becomes mediated through visual technologies (video, television, online). In one sense, it could be argued that performance magic, by its definition, has been leading the race when it comes to entertainment via illusion, while record production could only start partaking, after recorded sonic objects would become subjects to multitrack manipulation. But for both artforms, the handling of illusion reaches a 'meta' level with the adoption of, respectively, mediated technologies for magic and sampling technologies for record production; a comparison that uncovers important aesthetic issues in sample-based Hip Hop and explains some of the 'magical' analogies so frequently made about the artform.

One of the first problems affecting both artforms in terms of mediation is the issue of communicating performance authenticity, and the strategies that can ensure a convincing effect. Landman (2013, p. 60) asserts that conversely to televised magic, "live performance magic can develop experiences and feelings relating to trust and belief": participants in his magic workshops have reported feelings of increased trust due to the live nature of his performances, at the same time refusing to believe televised magic shows like the Derren Brown series. Leddington (2016, pp. 259–260) attributes the problem to an increase in physical distance between performer and audience, which becomes counterproductive to the experience of magic and constitutes an aesthetic issue that gets magnified with mediation: "The problem of distance is especially acute when showing magic on TV, where effective performance also requires ruling out the possibility of camera tricks and postproduction effects"—to that effect, TV magic shows like David Blaine's *Street Magic* deploy the portrayal of live audience reactions as stratagems "to certify the authenticity of the performance". Videos of magic tricks currently populating social media also deploy similar techniques, frequently staging participants around the performer in order to convince viewers of a certain degree of transparency.

To return to Reinhart's (2015, p. 35) analysis of magic portrayed in motion pictures, his concluding observation about *The Prestige* is that after the narrative has run its structural course of "pledge-turn-prestige", the film "introduces another meta-level by turning the cinematic narration into a magic trick by itself"; as a result "[w]e, the meta-audience, are tricked as well". The significance of this observation for sample-based practices is that it mirrors the meta-effect of being allowed, as a listener, in to the phonographic dimension of the sampling producer, witnessing the manipulation of previously made phonographic constructs. We may not be tricked, but we are entertained, and our interest is directed towards perceiving at least two, if not multiple, temporalities of phonographic process—in other words, we are hearing *process upon process.*

For a musical artform so heavily dependent upon music technology since its very inception, the problem of authenticity becomes magnified because of this degree of exponential mediation. Williams (2011) explains how recorded Hip Hop was initially perceived as an inauthentic take on the aural tradition of rap, especially when compared to the way it was being performed live in the Bronx in the 1970s. This perception was not helped by the fact that the first crossover hits credited as Rap had little to do with the music of the streets, something we have seen Kulkarni (2015, p. 37) describe as "essentially R'n'B records with rapping [and scratching] on them", to which he adds: "The crucially exciting thing about hip hop, the music made by scratch DJs, only figured as an effect, a detail, not the root of where the grooves and sounds came from".

Kulkarni here points to what may seem like a reversed notion of authenticity in Rap when compared to other musical genres: turntable performance using phonographic sources resonates from the origins of the culture, while live musicianship does not. Because of that, "rap music production… has aimed to create the sounds of the street", the strategy here consisting of the inclusion of "turntablistic codes on recordings" such as "vinyl scratching… to signify authenticity" (Williams, 2011, pp. 151–154). By the time sampling technology had become affordable, many of these turntablistic utterances were first replicated and later on developed further by DJs-turned-sampling-producers (something exemplified by the DJ prefix in many of the early hip-hop producers' aliases)—resulting in Hip Hop's Golden Age, boom-bap sound. As a result, the practice of sample-based Hip Hop developed its own intrinsic code of ethics regarding both sampling practices and what constitutes acceptable phonographic source material (Schloss, 2014). And while pop and rock record productions were creating 'supernatural' sonic constructs out of instrumental performances, sample-based hip-hop practice kept alluding to a performative tradition that represented the "alchemy" of creating "something out of nothing" (Kugelberg 2007, cited in Williams, 2011, p. 133); or, in other words, it alluded to exponential illusions conjured out of the flux of turntable-turned-sample-based manipulations over full phonographic sources. But let's look at the practice in more detail to investigate what constitutes beat-making 'alchemy' through a representative case study.

On track 'Musika' from KRS-One and Marley Marl's album *Hip Hop Lives* (2007), Marl samples the last few seconds of 'A Theme For L.A.'s Team', the opening track from motion picture soundtrack *The Fish That Saved Pittsburgh* (Thomas Bell Orchestra featuring Doc Severinsen, 1979). 'Musika' features reggaeton rapper *Magic* Juan and, to complete the metaphysical serendipity, the mystical theme is also present in the lyrics. The relevance of the example stems from Marl's celebrated status as an architect of the sample-based aesthetic in Hip Hop;[5] KRS-One's dedication to authentic hip-hop 'genre scripts' *and* the supernatural as a source of his inspiration;[6] but also because the track—and album—exemplify the practice of the Golden Age boom-bap aesthetic. In more detail, 'Musika' features about ten seconds from the 1979 recording, slightly sped up and looped into a cyclical structure, over which Marl builds a drum beat plus sparse additions of sub bass and a one-note repeated synthesiser bass figure taking place at the end of every four or eight bars. It is impossible to identify precisely the origin of the drum sounds contributing to the beat but, in the 'making-of' documentary included with the album release, KRS-One refers to Marl's pioneering practice of 'chopping' up individual drum hits from funk break-beats. Additionally, the timbre and dynamic envelopes of the sounds are characteristic of Marl's (and by extension Boom Bap's) practices of layering individual funk drum hits with synthetic 'boom' sounds from classic drum machines (such as the Roland TR-808). The soundtrack sampled has been mostly recorded at Sigma Sound (and other studios) but, judging from the timbral and spatial qualities of 'A Theme For L.A.'s Team', an educated guess would place the specific track at the famous Philadelphia location too (Nelson-Strauss, 2017). Toby Seay (2012) provides an illuminating rationale for the timbral and spatial characteristics that came to be known collectively as the 'Philly sound', and the sampled track in question subscribes to these. What is particularly telling about the Sigma sound is its unique echo-chamber footprint upon recordings actualised at the Philadelphia location, but also the rich, layered strings texture the personnel acquired by overdubbing the string section whilst inadvertently capturing speaker 'bleed' (Seay, 2012).

As a result, the original recording (which features eight individual instrumentalists alongside a string and horn section) carries with it a number of sonic illusions: layered instruments so they sound like larger sections; superimposed acoustic spaces (echo chamber) upon the actual spaces captured due to reflections during recording; re-amplified instrumental sections (and their reflections) captured due to 'spill' during overdubbing; as well as all the sonic artefacts and timbral processing colouration caused by recording and mixing practices, and the respective equipment used.[7] We might agree with Reynolds (2011, p. 313) that "[r]ecording is pretty freaky then", but let's explore why "sampling doubles its inherent supernaturalism". The section Marl uses clearly features a high-register trumpet solo, over a string section ostinato, but we can also hear the trademark Sigma Sound ambience. The strings are very rich in texture as a result of the overdubbing approach, occupying a wide stereo image and implied depth (illusion), which is typical of the Philly sound. Figure 4.2a provides a schematic representation of the sonic 'space' occupied by the sampled section.

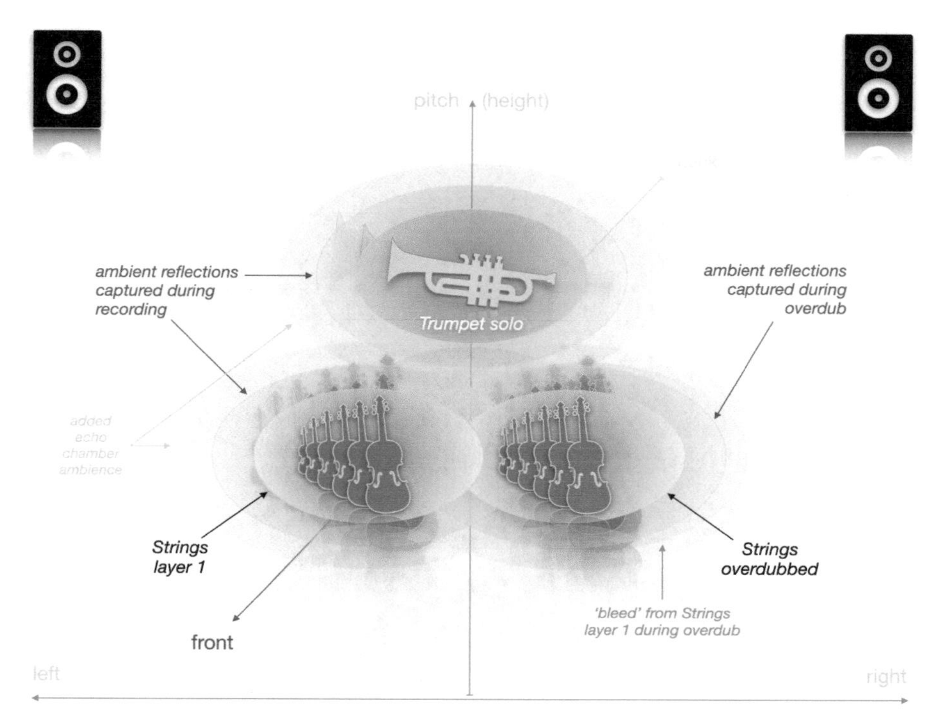

FIGURE 4.2a A schematic representation of the sonic 'space' occupied by the sample from 'A Theme For L.A.'s Team'. The visual representation of the sonic objects' pitch (frequency), stereo image, and depth in the mix is inspired by David Gibson's (2008) visual conceptualisation of mix layers, but also Moore and Dockwray's (2008) 'sound-box' illustrations.

Although Marl does not excessively manipulate or chop the original sample, the frequency content sounds higher than on the original track due to pitch-shifting (he could have achieved this by raising the cycles on a turntable prior to sampling the section or by tuning the sample higher within a sampler), but potentially also due to additional equalisation. Hip-hop producers will often manipulate the spectrum of a whole phonographic sample in order to make 'space' for the new elements they bring to the mix (including the rapper's voice). It is difficult to discern whether Marl has added any further reverberation to the sample, therefore superimposing yet another space upon the 1979 spatial illusions, but this—again—is common beat-making practice aiming to 'glue' all the borrowed elements within a new implied 'stage' (as we have seen in the practical expositions of Chapter 2). The low-frequency sounds (kick drum, sub bass, and bass synthesiser) come across as completely 'dry' (i.e. not carrying any substantial ambience) in the hip-hop mix, which places them rather 'forward' in the staging illusion. The drum sounds (gathered from a multitude of sources) typically feature the characteristics of both 1970s drum sources and 1980–1990s hip-hop drum layers, but their truncation and any dynamic envelope-shaping pack the contained ambiences into unnaturally abrupt durations. Figure 4.2b provides a schematic representation of the resulting staging illusions in 'Musika'.

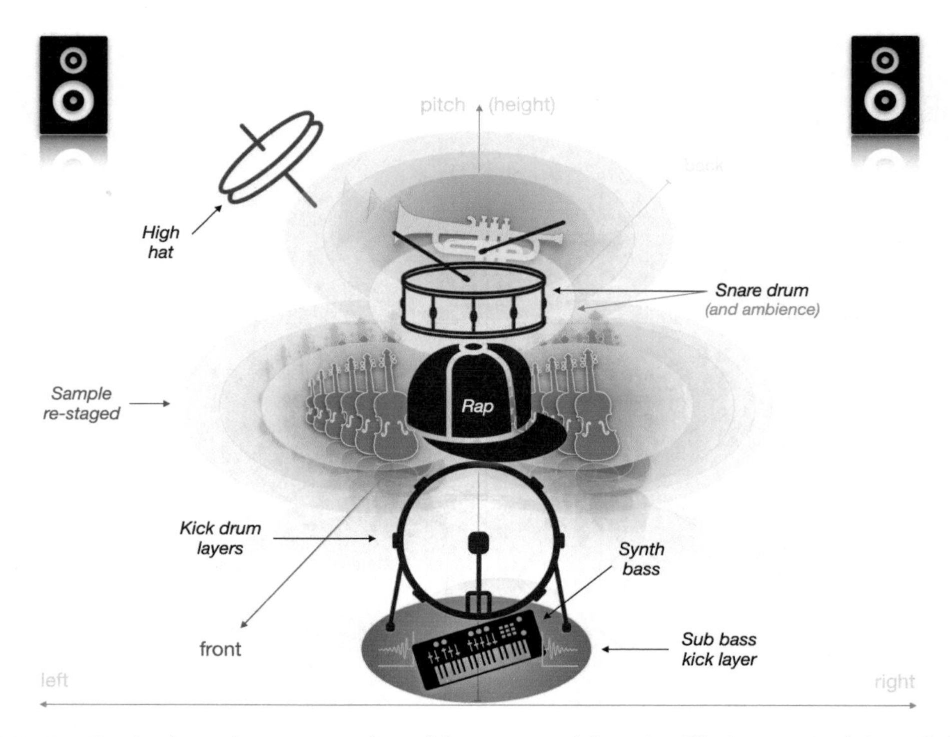

FIGURE 4.2b A schematic representation of the exponential staging illusions on track 'Musika'.

The example illustrates that what may sound motivically quite repetitive and simple on the surface, represents, in fact, a complex, multidimensional, and rich *sonic* construct. What is noteworthy here is that the juxtaposition of the plethora of timbral, dynamic, and spatial illusions creates an exponential one, which is, however, held together by the sample-based producer's craft (control). The sampling producer is drawn to the phonographic sources because of the rich sonic phenomena they contain, but also because the origins of the artform predispose the beat-maker to 'jamming' with the past. The "sample collage" as a "musical event that never happened" may indeed be "the musical art of ghost co-ordination and ghost arrangement" (Reynolds, 2011, pp. 313–314), but unlike Reynolds's dissatisfaction with the artform, the listener of sample-based Hip Hop remains engaged and entertained; and this is due to a stylistic contract that presumes suspended disbelief conditioned by the evolution of record production. The 'alchemy' or command that the producer demonstrates over the multiple sonic dimensions may rightfully sound 'magical', but it is exercised through the craft of rhythmic, dynamic, and timbral organisation—Krims's (2000) "hip hop sublime"—having first tamed the tools of the trade: the relevant sampling technology.

Impossibility, motion, and the ecology of 'alief'

In his article 'The experience of magic', Jason Leddington (2016, p. 254) sets out "to initiate a philosophical investigation of the experience of magic with a focus on its cognitive dimension" and sees his work as "a first step toward a general aesthetics of the impossible". His examples for "impossible music" include Risset rhythms and Shepard tones (Leddington,

2016, p. 254), which "can be constructed to give the perception of continuous acceleration" (Stowell, 2011, p. 1) or pitch oscillation, respectively. But this kind of "fractal self-similarity" (Stowell, 2011, p. 1)—entertaining and extreme as it is—is not the only kind of sonified 'impossibility' we may get exposed to as we have seen. Figure 4.2b represents schematically, a sonic experience that is more comparable to Reutersvärd's 'Impossible Triangle' or Escher's 'Irrational Cube', to use two of Leddington's visual examples. We can clearly see these shapes on two-dimensional paper (and they would not be too difficult—and certainly not impossible—to recreate by drawing); but if we try to imagine them in 3D space, our perception reaches a moment of cognitive dissonance: the shapes can be drawn, but what they represent in three dimensions cannot exist in physical space.

Similarly, the exponential supernaturalism of the sample-based music collage presents not only spatial but also temporal impossibilities—musicians and sonics from different eras co-existing on multiple spatial planes. Here, we perhaps reach an irrefutable analogy between recorded music and magic. "Magic is... about creating... the illusion of impossibility" (Ortiz, 2006, cited in Leddington, 2016, p. 254), just as it is for impossible/irrational shapes, for phonographic records following the advent of multitracking, and—exponentially so—for hip-hop productions making use of sample-based practices. There appears to be a universal magnetism drawing humanity towards experiences of awe that cannot be (easily) explained. On the surface, this seems like an oxymoron—being drawn to self-inflicted illusions of impossibility and moments of cognitive dissonance, or choosing to be entertained by the "bafflement of (our) intellect" (Leddington, 2016, p. 259). Could this be a form of acknowledgement of other modes of knowledge, or of the limits of the current state of affairs in rational and scientific thought? Leddington (2016, p. 264) reminds us that "Socrates's 'human wisdom', which consists in his knowing only that he does not know, is a form of sustained *aporia*" (my emphasis); and "maintaining the belief that there is a correct account of piety (or virtue, or justice,...) in the face of *aporia* is of paramount *ethical* importance" (original emphasis). This notion is echoed across the scientific world, but also in studies investigating the evolutionary purpose of music. In his investigation of the overlap between magic and science, Reinhart (2015, p. 34) draws our attention to the fact that "Einstein is aware that a pure positivist view can't possibly cover the full extent of what is comprehensible by the human mind". Perlovsky (2017, pp. 28–31), furthermore, traces this conflict down to the cognitive mechanisms of differentiation and synthesis, championed respectively by the development of language and music:

> Most of the knowledge that exists in culture and expressed in language is not connected emotionally to human instinctual needs... While language splits psyche, music restores its unity. We come to understanding why music has such power over us: we live in the ocean of grief created by cognitive dissonances...; and music helps us alleviate this pain.

Leddington's theorising leads him to a further powerful argument. The commonly used notion of (willing or unwilling) *suspended disbelief* cannot sufficiently explain the enjoyment we experience from engaging with illusions of impossibility. If it were so, how could we actually be amazed (entertained)? He believes, instead, that the necessary condition must be to experience—even to maximise—"cognitive dissonance that is not a matter of conflicting beliefs" (Leddington, 2016, p. 257). To explain the cognitive mechanism, Leddington borrows the notion of *alief* from Szabó Gendler (2008, cited in Leddington, 2016, p. 257), defined as follows:

> A paradigmatic alief is a mental state with associatively linked content that is representational, affective and behavioral, and that is activated—consciously or nonconsciously—by features of the subject's internal or ambient environment.

Gendler (cited in Leddington, 2016, p. 257) goes on to explain how a conflicting experience such as walking on the transparent Grand Canyon Skywalk bridge would involve a clash of an intellectual belief in presumed safety "and a more primitive, nondoxastic, representational mental state she calls *alief*" (original emphasis). Anyone with a fear of heights would have experienced a similar conflict between, on the one hand, the intellectual reassurance of a situation as safe and, conversely, an irrational fear about approaching a barrier or looking down. Ortiz (1955/2011, cited in Leddington, 2016, p. 258) pins down the recipe for successful magic in this very tension, expressed as the victory of emotional over intellectual belief. He offers a telling 19th-century anecdote to illustrate their difference: "Madam De Duffand was asked whether she believed in ghosts. She responded, 'No. But I am afraid of them'" (Ortiz, 1955/2011, cited in Leddington, 2016, p. 258). If we take a recent sample-based hip-hop record such as 'The Story of O.J.' (from *4:44*, 2017) by Jay-Z—produced by No I.D. and featuring samples of 'Four Women' (from *The Best of Nina Simone*, 1969)—the equivalent exchange between two fans might sound something like this:

FAN A—Do you believe Jay and Nina performed on this together?
FAN B—No. But their interactions *move* me.

In his theory of an ecological approach to the perception of musical meaning, Clarke (2005) echoes Gendler's notion of alief as a state activated by features of the environment. He suggests that "perception must be understood as a relationship between environmentally available information and the capacities, sensitivities and interests of a perceiver" (Clarke, 2005, p. 91):

> An important component of that subjective engagement with music is its corporeal, proprioceptive, and motional quality, which may on occasion provide listeners with experiences of "impossible worlds" that have some of the same attractions as do other forms of virtual reality.
>
> *(Clarke, 2005, p. 90)*

Levitin (2006, p. 192), furthermore, supports a primordial rationale behind our instinctive, embodied response (motion) to strong rhythmical content present in a musical track, which is congruent with a notion of being coerced to move (dance, nod, tap our foot) by the (recorded) commands of a music producer, despite any intellectual identification of temporal or spatial sonic 'impossibilities':

> Our response to groove is largely pre– or unconscious because it goes through the cerebellum rather than the frontal lobes... [Your brain] involves a precision choreography of neurochemical release and uptake between logical prediction systems and emotional reward systems.

No I.D.'s highly rhythmical chopping of Nina Simone's 'Four Women' against his programmed beats on 'The Story of O.J.' therefore *move* us (emotionally and arguably physically) despite: the impossible (or non-natural) resulting vocal phrases; the juxtaposition of Nina Simone's and Jay-Z's voices; and their different but characteristic phonographic signatures (signifying both 1960s and 2010s recording aesthetics—themselves the result of different production practices and equipment/media used). Interestingly enough, the two characters are brought together in the song's music video through animation, concurring with Clarke's (2005, p. 86) "impossible world" analogies across different artforms.

Freud's (1975/2001, p. 90) position then that only in art and magic can mimetic action be thought to influence recipients and produce emotional effects "just as though it were something real" rings true. But it is not the result of blind confidence in the performer's power of control, but more so a transference of action upon artistic materials, themselves in turn communicating 'instructions' embedded in the work. These can psychologically *move* the recipient, and in the case of rhythmical transference in music, *literally* move the listener. Hazrat Inayat Khan (cited in Godwin, 1987, pp. 261–262) makes a telling leap from the metaphysical to the physical effects of sound:

> [E]very sound made or word spoken before an object has charged that object with a certain magnetism... The whole mechanism, the muscles, the blood circulation, the nerves, are all moved by the power of vibration. As there is resonance for every sound, so the human body is a living resonator for sound ... Sound has an effect on each atom of the body, for each atom resounds.

Zagorski-Thomas (2018, pp. 347–348) supports the notion with a more scientific perspective:

> A crucial piece of information from neuroscience is that we recognise human gesture by mentally "doing it" ourselves ... If our interpretation of the world through firsthand experience is schematic in nature, the way we create meaningful, symbolic representations of aspects of our experience through language, gesture, and the manipulation of our environment takes this schematic nature to another level.

Because the performer/producer is attempting to create a finely balanced experience of conditions or, using Clarke's approach, an environment of engaging information echoing nature to present to a perceiver (which in Clarke's interpretation also includes culture), the achievement of such an architecture, sonic 'world', or effective illusion is referred to as magical. The inexplicable, the 'bafflement of the intellect', and the resulting awe are effects borne out of resonating with a humanly constructed 'sublime environment'. And the effect is not just a perceiver's gift; the creator can also be mesmerised by their own achievement because achieving control over the infinite variables is not a given, but a harmonious plateau reached when the performer's level of skill and the environmental variables meet. It is what externally appears as control, what artists experience as "flow" (Csikszentmihalyi, 1990), the results characterised as 'sublime'.

In sample-based music creation, the infinite network of sonic possibilities includes previous phonography. Although practitioner mastery is a condition, "flow" is reached only when control and environmental challenge reach a balance (Csikszentmihalyi, 1990). The dynamic is perceivable by both makers and recipients, and this is why consistent practitioners acquire magical characterisations from their community (peers, fans); why the construction of 'sublime' sonic worlds is called magical; and why practitioners, at times, refer to their own process *as if by magic*. It can't be so but it is. Both real and unreal. Coercing the listener into sympathetic motion through constructed experiences that trigger the right kind of neural mirroring (Cook et al., 2014), as Zagorski-Thomas demonstrates.

When asked about his favourite creation on his debut mixtape *Griselda Ghost* (Westside Gunn x Conway, 2016), mysterious rap critic turned prolific sample-based producer Big Ghost Ltd. (who, incidentally, never reveals his true identity) attests: "My absolute favorite beat even before they recorded any vocals to 'em was Fendi Seats. They [rappers Westside Gunn & Conway] also happened to both snap on that shit. That one came together like magic, B" (Shabazz, 2015). This is a typical acknowledgement of musical factors external to the practitioner's control. The sample-based producer demonstrates a default humbleness due to their dependence on a field of previous phonography, but also a notion of self-effacement founded upon aporia. Zak (2001, pp. 195–196) tells us that "[b]oth the artist's expressive gesture and the listener's interpretation are infused with an awareness of field that allows minimal, momentary, and inexplicable allusions, references, and rhetorical gambits to resonate in a frame far larger than themselves". This explanation highlights the forces of chaos involved in record production as a result of the plethora of sonic variables, technological options, and (sub)cultural associations available. And these become exponential in a form of sample-based phonography that itself interacts with *previous* phonography. The phonographic moments that succeed in aligning these 'inexplicable allusions' therefore feel magical to both makers and listeners, because they feel rare; because they feel sublime; because they feel *impossible*; because we find their illusions of impossibility not only entertaining, but also healing for our intellect; and because they provide *synthesis*.

But how is technology itself capable of leading to further magical manifestations? Likening the computerised and networked present to a magical world, Wilcock (2015, pp. 43–44) warns about the "ability to take control of someone's soul" with current technology, if soul is defined as "all the information about one's self, all the information that makes up one's online presence/self, or 'the algorithm' that summarised all that you are". The analogy to a stored instance of someone's voice within a sample (e.g. Simone's presence in 'The Story of O.J.') is an easy leap to make. The mapping of vocal phrases on the drum pads of a sampling drum machine leads to rhythmical and textural manipulations enforced upon past voices/performers/sounds; and these can be seen as a form of 'phonographic conjuring' and, in turn, a kinaesthetic (embodied) coercing of the listener (based on the neural mirroring effects discussed above).

The temporal distance perceived between the percussive actions of the contemporary producer (No I.D.'s reimagined patterns triggering Simone's voice) and the performer situated in the past (Simone's 1969 performance) becomes a further condition for a magical experience. Even though distance between performers and audiences is counterproductive to magic as Leddington (2016) has shown, distance between user and source is in fact a condition for successful magic to occur according to the "law of contagion" (Wilcock, 2015, p. 40): "things which have once been in contact with each other continue to act on each other at a distance after the physical contact has been severed" (Frazer cited in Wilcock, 2015, p. 49). Hearing the interaction of old and new musical or sonic utterances is contagion manifested on rap records. Figure 4.3 illustrates the drum pads of an Akai MPC sampling drum machine 'loaded' with samples summarising the phonographic examples discussed in this chapter.

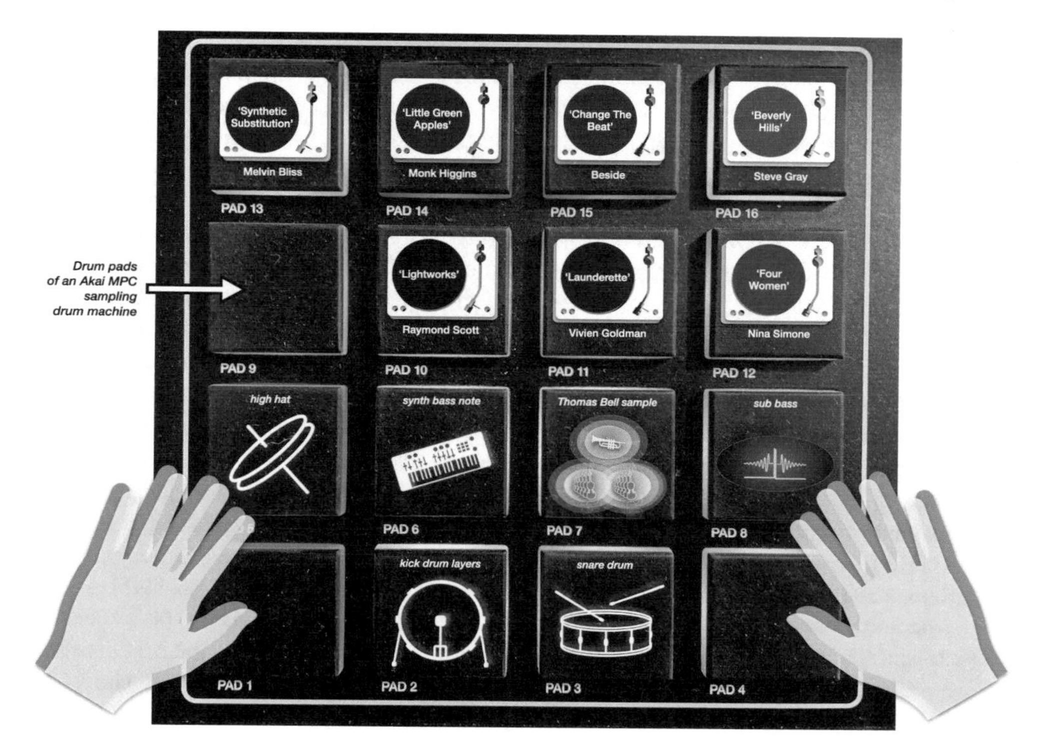

FIGURE 4.3 A schematic representation of the sampled examples discussed in this chapter on the drum pads of an Akai MPC sampling drum machine.

In practice

There are a number of beats in the associated album that pursue and manifest aspects tied to the notion of 'sample magic': these include the design of source constructs fertile in sonic illusions that inspire a sense of performative mediation in the sample-based praxis; followed by a conjuring kind of gestural 'play' with their sonic manifestations through forms of control exercised in the beat-making process itself. Some of these illusions include stylistically informed 'aged' timbres, as well as consciously constructed poly-dimensional mix architectures, providing the 'magical' ingredients towards phonographic revenants that can then be animated in the sample-based 'theatre of play'. The following examples illustrate.

Example 1

'Sub Conscious' features a source mix that was put together as a rather spontaneous jam by overdubbing parts on an upright piano, electric Rhodes piano, and fretless bass—not dissimilar to the concept behind Bill Evans's *Conversations with Myself* (1963), in which he layers multiple piano parts reinterpreting jazz standards and original composition.[8] The catalyst, however, behind the spark of the recording session had been time-limited access to a stereo pair of prized microphones—modelled on vintage classics—available to me only for one short evening. Placing them on the far sides of the open upright piano, I 'made sound' by inventing a simple, soulful harmonic sequence and improvising around it in sync to a metronome track. I 'fed' the microphones a spectrum of notes and clusters in various registers (following the harmonic movement), testing the microphones' response and getting inspired by the foldback mix that was coming back to me via the headphones. After a few minutes of play, I moved to the Rhodes piano, 'responding' to the upright piano part already captured. The fretless bass layer provided a gentle foundation under the unfolding 'conversation', rooting the keyboard dialogue. I made sure the electric instruments were played through amplifiers, which I recorded with microphones to take advantage of a retrospective "symbiotic" relationship between them and the acoustic piano—a shared co-existence of their "harmonics in space" (Dorrough cited in Horning, 2013). I felt a sense of 'flow'—everything happening very fast, with minimal effort, time seemingly collapsing. I would normally allow for some distance between recording, mixing, and sampling sessions, but I was mesmerised, so I quickly dialled up emulations of vintage tape machines, pre-amplifiers, mixing desks, and spatial effects on the computer, and with minor adjustments, I enhanced the symbiotic 'glue' already captured via shared post-production decisions. The output of the computer mix had been permanently pre-wired to the inputs of two samplers connected in series for this project (a Roland SP-404 first, into an Akai MPC X second), to take advantage of their different sonic footprints, and—at times—to manipulate sound in one, before capturing 'chops' in the other. Late into the early hours, I found I had a series of chops looping around in what felt like a haunting sequence and—inspired by the more recent trends in New York sample-based Hip Hop—I proceeded to add some single boom-bap drum hits, but in a subtle, evolving way in the arrangement. I reinforced the kicks with long 808 sub bass notes that I tuned to the roots of the fretless bass, but left some spaces for the Rhodes to breathe, and then progressively distorted the 808s to create a dynamic variation through a more emphatic B-section. Samples of real vinyl crackle (from the end of LPs) were overlayed in a rhythmical sense over the stereo mix/master chops, giving them surface texture, and I reached out for short, ornamental vocal/lyric phrases from a series of prior, blues-inspired recording sessions to accentuate the mobilised sonic worlds/objects with narrative 'echoes from the past'. This was probably the fasted overall production journey of any track on this album.

The 'vignette' above illustrates the contextual factors that framed a phonographic occurrence, inspirational to the beat-making journey, through the sum of spatial and timbral factors captured in the recording and mixing processes. The sample-based composition manifests control over the inspiring sources both rhythmically (and harmonically), interacting with the chopped utterances through the initially subtle beat and sub-bass programming, but also in terms of the variety of congruent timbres selected and sculpted: the manipulation of the 808 textures is balanced against the fretless bass spectra, and the new drum hits are chosen in a cyclic jam against the rhythmical occurrences resulting from the interaction of chopping decisions and syncopation inherent within the tracked keyboard and bass dialogue(s).

The blues voice signatures, providing emphasis over the instrumental movement, are further distanced in time (and space) via additional filtering and spatial processing. Expressing the 'law of contagion' through timbral-spatial distancing is particularly useful in this case, as the voice is my own, and turning it into a 'past' character helps mobilise it as yet an/other element of the unfolding sonic script. This kind of gestural control is exercised over numerous short vocal elements performed by me, stemming from various stylistically-infused source production phases. Consequently, this offers a unifying narrative thread—a kind of timbral leitmotif—across the associated album. For example, aggressive vocal segments and shouts from the alt-rock, punk, and hardcore content provide emphasis and rhythmical interplay on beats such as 'Altcore', 'Become', 'Good Ole Betty', 'Oh Buxom Betty', 'Train Brain', and 'Covert Three' (mostly in Greek); while more of the blues-infused segments that acted as rhythmical ornamentation and emphasis on 'Sub Conscious' extend the 'blues narrator' function over beats such as 'Alderoots', 'Americanotha', 'Call', 'It Meters', 'King's Funk', 'Mo Town', 'Reggae Rock', 'Studio A', and 'Tense Minor' (mostly in English—though in both cases lyrical intelligibility is a secondary function to surface ornamentation and rhythmic/emotive emphasis).

Example 2

Furthermore, the performative-layering and mixing approaches that shaped 'Sub Conscious' are echoed in constructs from all stylistic eras simulated in the source production process. The following journal entry provides an example from the reggae/dub phase, highlighting the interaction between tracking improvisation and post-production decisions that led to samplable moments:

'Nu Drub'

As I was looking for the right bars to sample from one of my dub experiments, I realised that beyond the effectiveness of the musical 'conversation' between my different overdubbed layers, the moments that really grabbed me were those when the interaction became sonic, not just notational or rhythmical. After about 120 bars of improvisation, perhaps the right brain kicked in, locking in to the less numerical or obvious aspects of the jam. There was space. There was textural complementarity, there were rhythmical, motivic, *and* sonic *'interactions'* in the overdubbed layers: tones drawn from the instruments; spatial and tonal effects dialled in on the pedals and tracking equipment; and a more (sub?)consciously 'aware' monitoring of the 'ghosts' beyond the notes—echo and reverb tails, and what should be played against them. (original emphasis)

When it came to mixing this, leaving the automated responses behind, I realised that what were working as 'phonographic' moments worth sampling, had been achieved through this tracking synergy, as well as four post-production 'enhancements': a(n AKG BX-20) spring reverb; a(n EP-34 Echoplex) tape delay (both shared via the sends of multiple channels); a (Neve 33609) mix-buss compressor; and a(n Ampex ATR-102) master tape machine (all emulations). I decided not to over-process, go with the flow, and remind myself [that] we sample great records, not necessarily technically perfect mixes.

Source constructs stylistically similar to these, rich in spatial depth and timbral character, have ignited a series of beat-making experiments leading to further sample-based productions such as 'Dragga Five', 'Old Steppers', 'Reggae Rock', and 'Steppers Dub'—the titles indicating their (inter-)stylistic referentiality to source objects used in their inception.

Example 3

From a recording perspective, the spatial illusions that enable the interplay with, gestural control over, and juxtaposition of poly-dimensional sonic worlds in the beat-making stages extend to tracking decisions. Acting in an engineering capacity whilst recording two songs for groups Sarabanda and Grupo Lokito, I was aiming at a sound that blended vintage and modern phonographic signatures in the respective traditions of Latin and Afro-Cuban record-making. The objective had been to capture the characteristic performative 'liveness' these bands are known for, whilst ensuring a level of control over individual instrumental sonics, helpful to the post-production process. The approach is described in the following vignette, echoing the pursuit of 'symbiotic harmonics in space' as in the first example, but with a mixture of simultaneous and overdubbed tactics:

After discussing and playing a number of Latin and Afro-Cuban phonographic references with Sara [McGuinness]—the musical director, pianist, and arranger of both bands—we formulated a recording plan. We wanted to record all members of the core rhythm section simultaneously (congas, drums, electric bass, and piano), ensuring visual and sonic communication for the members, but separating the percussive players from the bassist and pianist. This allowed a certain degree of control. I placed the conga player and drummer at two opposite corners of the live room, and directly connected the bassist and pianist (using a keyboard, initially, for the piano sound) in the control room where I was operating the mixing console. The distance between the percussionist and drummer minimised direct phase issues, but I also wanted to creatively utilise 'spill' between the two instruments on each other's microphones. I additionally wanted to 'print' the sound of the [live] room along with the sources to maximise the illusion of all members sharing a space in the mix. So, inspired by Stavrou's (2003, pp. 47–49) "Turn Spill into Ambience" technique, I close-miked the congas and elements of the drums, but also set up pairs of farther, overhead microphones over both instruments to capture them with added room reflections.[9] Using headphones in the control room set to mono playback, I made sure that the vintage overhead mics I positioned over the congas (set to a cardioid polar pattern) were in phase with each other, the close microphones on the congas, as well as the drum microphones behind them. This took some repositioning until all combinations of microphone perspectives sounded reinforced by each other [symbiotic and in phase]. After a successful performance of the rhythm quartet, the conga player overdubbed bongos and campana (cowbell) in the same position as before (and with the same microphone combination). We muted the guide keyboard part, and Sara reperformed it on the actual baby grand piano positioned in the live room. I redeployed the vintage microphones previously used as overheads on the congas and bongos to track the piano, aiming for a balance of direct and room sound, by positioning them at some distance from the piano's sound board. On a different date and in a different studio, we recorded additional percussion, guitar, keyboards, horns, and the vocalists using similar microphone setups. I mixed the songs on a large-format hardware console, taking advantage of the analogue summing, and using reverb processors to 'marry' elements from the two different recording environments into virtual shared spaces, emulating rooms with complimentary qualities and dimensions to the real ones captured. I printed both the complete mixes, as well as stereo stems of instrumental groups alongside their surrounding environments, for future access. The complete stereo mixes were professionally mastered for release.

The two beats made out of these songs are 'Como Mi Ritmo' and 'La Noche (AMHB)', with the former illustrating a chopping process utilising the full stereo master, and the latter deploying a chopping approach over the instrumental stems. The end results are telling of the level of access to the source material, with 'Como Mi Ritmo' gesturally mobilising short, crystallised objects from the song's master, and 'La Noche (AMHB)' interweaving instrumental and vocal segments into a more complex tapestry resembling a remix. The horns from 'La Noche (AMHB)' also appear recontextualised—but incorporating their original timbral and spatial footprints—in 'Mo Town' and 'Rebluezin'.

Conclusion

The mapping of parallels between two artforms can often appear as nothing more than an intellectual puzzle, entertaining the scholar with fanciful surface analogies but doing little to uncover any potent aesthetic insights. It is my hope that by investigating the frequent associations made by practitioners, fans, critics, and scholars between magic and sample-based music, this chapter has demonstrated the rationale behind such analogies and, furthermore, it has started to uncover a number of parallel mechanics lying under the surface of the two artforms. Specifically for Hip Hop, the invested interest lies in understanding the appeal of the sample-based aesthetic in a number of dimensions: the lure of phonographic samples for practitioners, but also the magic of the sample-based aesthetic for listeners. Furthermore, as the chapter has shown, the evolving nature (and exponential hybridisation) of subgenres promotes new questions about the practices and emerging aesthetic issues to the forefront, and the essence of 'magic' in sample-based processes acquires increased urgency. By studying the dynamics of this interaction between raw sonic materials and sampling in the established practice of utilising phonographic sources, (we) practitioners are empowered to infuse the 'magical' qualities necessary into both (new) sources and process, facilitating the future development of the genre. This becomes the extrapolation of the following two chapters, first examining the notion of infusing unique, magical *alterity* into newly constructed sonic objects, followed by investigating the mechanics of their juxtaposition in this alternative sample-based collage proposition.

Recommended chapter playlist
(in order of appearance in the text)

'Sub Conscious'
'Altcore'
'Become'
'Good Ole Betty'
'Oh Buxom Betty'
'Train Brain'
'Covert Three'
'Alderoots'
'Americanotha'
'Call'
'It Meters'
'King's Funk'
'Mo' Town'
'Reggae Rock'
'Studio A'
'Tense Minor'
'Nu Drub'
'Dragga Five'
'Old Steppers'
'Steppers Dub'
'Como Mi Ritmo'
'La Noche (AMHB)'
'Rebluezin'

Notes

1 For a complete account on the development of turntable practice—or 'turntablism'—as instrumental practice, see Katz (2012, pp. 43–69).
2 This is a notion that echoes Ihde's (2012, p. 108) "fantasy variations": "It is out of possibility that the undiscovered is found and created".
3 It would be hard to imagine how, for example, the renaissance, a shift towards scientific thinking, technological awareness, and secularism in the West would not have affected audience trends in their engagement with magic.
4 Moore (2002, p. 213) discusses this form of authenticity as "primality".
5 Rose (1994, p. 79) informs us that: "A few years after rap's recording history began, pioneering rap producer DJ Marley Marl discovered that real drum sounds could be used in place of simulated drum sounds".
6 KRS-One published a hip-hop 'commandments'-style/dogma book entitled *The gospel of Hip Hop: First instrument* (2009).
7 Moore (2019, p. 210) defines audio colouration "as the subtle, and sometimes not so subtle, changes in program material that manifest perceptually as variations in timbre"; which "occurs when audio equipment alters features of the original program material, including (but not limited to) changes in the frequency response, dynamic envelope, and harmonic content, through the addition of harmonic components…".
8 Jazz critic Scott Yanow (no date) describes the album in the following words: "Aptly titled, the music on this LP has a surprising amount of spontaneity, with Evans constantly reacting to what he had just recorded, and the results are sometimes *haunting*" (my emphasis).
9 Stavrou (2003, pp. 47–48) describes how he turned the compromise of 'spill' between double bass and horns—sharing a space on Vince Jones and Grace Knight's 'Come In Spinner' (1990)—into a complimentary ambient characteristic:

> Imagine a blaring brass section next to the gentle tones of an upright double bass. Whatever you do, you're going to get spill. This could be a nightmare. You could box the double bass player in with gobos, but you'll destroy the performance ... [Instead] reverse engineer the bass sound. First, find a good spot for the brass ambience mic ... Now find a piece of air that contains good brass ambience, hopefully near where the double bass player feels comfortable playing ... Place the microphone such that the brass ambience enters it from the rear of the cardioid pattern ... Because the brass ambience is entering the rear of the cardioid pattern, it will be dull. This provides you with a good excuse to add some treble to balance the ambience, which also helps the double bass sound nice and clear. To put it another way, the EQ you use to brighten up the double bass also enhances the brass ambience, rather than exaggerate brass spill. You've turned a liability into an asset!

Bibliodiscography

Beside (1982) *Change the Beat* [Vinyl, 12]. US: Celluloid.

Bliss, M. (1973) *Reward / Synthetic Substitution* [Vinyl, 7]. US: Sunburst Records.

Caldwell, B. (2015) *The Awesome B.B. King/UGK Mashup, Houston Press.* Available at: https://www.houstonpress.com/music/ugk-bb-king-mashup-the-trill-is-gone-is-as-awesome-as-you-think-it-is-7887211 (Accessed: 1 December 2020).

Clarke, E.F. (2005) *Ways of listening: An ecological approach to the perception of musical meaning.* Oxford: Oxford University Press.

Cook, R., et al. (2014) 'Mirror neurons: From origin to function', *Behavioral and Brain Sciences*, 37(2), pp. 177–192.

Csikszentmihalyi, M. (1990) *Flow: The psychology of optimal experience.* New York: Harper & Row.

Evans, B. (1963) *Conversations With Myself* [Vinyl LP]. UK: Verve Records.

Freud, S. (2001) *Totem and taboo.* Oxon: Routledge Classics.

Gang Starr (1994) *Hard to Earn* [CD, Album]. Europe: Chrysalis, ERG.

Gang Starr (2003) *The Ownerz* [CD, Enhanced Album]. US: Virgin.

Gibson, D. (2008) *The art of mixing: A visual guide to recording, engineering and production.* 2nd edn. Boston, MA: Course Technology.

Godwin, J. (1987) *Music, mysticism and magic: A sourcebook.* London: Penguin.

Goldman, V. (2016) *Resolutionary* [CD, Album]. Germany: Staubgold.

Higgins, M. (1968) *Extra Soul Perception* [Vinyl LP]. US: Solid State Records.

Horning, S.S. (2013) *Chasing sound: Technology, culture, and the art of studio recording from Edison to the LP.* Baltimore, MD: JHU Press.

Ihde, D. (2012) *Experimental phenomenology: Multistabilities.* 2nd edn. Albany: State University of New York.

J Dilla (2006) *Donuts* [CD, Album]. US: Stones Throw Records.

Jarrett, M. (2014) *Producing country: The inside story of the great recordings.* Middletown, CT: Wesleyan University Press.

Jay-Z (2017) *The Story of O.J., 4:44* [Digital Release, Single]. Roc Nation.

Jones, V. and Knight, G. (1990) *Come in Spinner* [CD, Album]. Australia: ABC Records.

Katz, M. (2010) *Capturing sound: How technology has changed music.* Berkeley: University of California Press.

Katz, M. (2012) *Groove music: The art and culture of the hip-hop DJ.* New York: Oxford University Press.

Krims, A. (2000) *Rap music and the poetics of identity.* Cambridge: Cambridge University Press.

KRS-One (2009) *The gospel of hip hop: First instrument.* New York: Powerhouse Books.

KRS-One and Marley Marl (2007) *Hip Hop Lives* [CD, Album]. US: Koch Records.

Kugelberg, J. (2007) *Born in the Bronx: A visual record of the early days of Hip Hop.* New York: Rizzoli.

Kulkarni, N. (2015) *The periodic table of Hip Hop.* London: Random House.

Lacasse, S. (2000) *'Listen to my voice': The evocative power of vocal staging in recorded rock music and other forms of vocal expression.* Unpublished PhD thesis. University of Liverpool.

Lamont, P. and Wiseman, R. (1999) *Magic in Theory: An introduction to the theoretical and psychological elements of conjuring.* Hatfield: University of Hertfordshire Press.

Landman, T. (2013) 'Framing performance magic: The role of contract, discourse and effect', *Journal of Performance Magic*, 1(1), pp. 47–68.

Leddington, J. (2016) 'The experience of magic', *The Journal of Aesthetics and Art Criticism*, 74(3), pp. 253–264.

Leddington, J. (2017) 'The enjoyment of negative emotions in the experience of magic', *Behavioral and Brain Sciences*, 40, pp. 34–35.

Levitin, D.J. (2006) *This is your brain on music: The science of a human obsession*. London: Penguin.

Lewis, G.E. (2017) 'Improvised music after 1950: Afrological and eurological perspectives', in C. Cox and D. Warner (eds) *Audio culture: Readings in modern music*. 2nd edn. New York: Bloomsbury Academic, pp. 385–398.

Liu-Rosenbaum, A. (2012) 'The meaning in the mix: Tracing a sonic narrative in 'When The Levee Breaks'', *Journal on the Art of Record Production*, 7.

Madlib the Beat Konducta (2006) *Vol. 1–2: Movie Scenes* [CD, Album]. US: Stones Throw Records.

Magic | Definition of Magic by Oxford Dictionary on Lexico.com (2018) *Lexico Dictionaries*. Available at: https://www.lexico.com/definition/magic (Accessed: 4 March 2018).

Moore, A. (2002) 'Authenticity as authentication', *Popular Music*, 21(2), pp. 209–223.

Moore, A. (2019) 'Tracking with processing and coloring as you go', in R. Hepworth-Sawyer, J. Hodgson, and M. Marrington (eds) *Producing music*. New York: Routledge (Perspectives on Music Production), pp. 209–226.

Moore, A.F. and Dockwray, R. (2008) 'The establishment of the virtual performance space in rock', *Twentieth-Century Music*, 5(2), pp. 219–241.

Nelson-Strauss, B. (2017) 'The Fish That Saved Pittsburgh', *BLACK GROOVES*, 4 April. Available at: https://blackgrooves.org/the-fish-that-saved-pittsburgh/ (Accessed: 27 February 2018).

Perlovsky, L. (2017) *Music, passion, and cognitive function*. London: Academic Press.

Reinhart, M. (2015) 'Spirited away', *Journal of Performance Magic*, 3(1), pp. 25–35.

Reynolds, S. (2011) *Retromania: Pop culture's addiction to its own past*. New York: Faber and Faber.

Rose, T. (1994) *Black noise: Rap music and black culture in contemporary America*. Hanover, NH: University Press of New England (Music/Culture).

Schloss, J.G. (2014) *Making beats: The art of sample-based Hip-Hop*. Middletown, CT: Wesleyan University Press (Music/Culture).

Scott, R. (2000) *Manhattan Research Inc.* [2xCD, Album]. Netherlands: Basta.

Seay, T. (2012) 'Capturing that Philadelphia sound: A technical exploration of Sigma Sound Studios', *Journal on the Art of Record Production*, 6.

Shabazz, S. (2015) 'A Conversation with Big Ghost | The Real Hip-Hop', 21 September. Available at: https://therealhip-hop.com/a-conversation-with-big-ghost/ (Accessed: 28 August 2021).

Simone, N. (1969) *Four Women, The Best of Nina Simone* [Vinyl LP, Compilation]. US: Philips.

Stavrou, M.P. (2003) *Mixing with your mind: Closely guarded secrets of sound balance engineering*. Australia: Flux Research.

Stowell, D. (2011) 'Scheduling and composing with Risset eternal accelerando rhythms', in *Proceedings of the International Computer Music Conference 2011*. Huddersfield: University of Huddersfield, pp. 474–477.

Thomas Bell Orchestra featuring Doc Severinsen (1979) *A Theme For L.A.'s Team, The Fish That Saved Pittsburgh* [CD, Album]. US: Lorimar Records.

Various (1979) *Relax* [Vinyl LP]. UK: Bruton Music.

Vendryes, T. (2015) 'Versions, dubs and riddims: Dub and the transient dynamics of Jamaican music', *Dancecult: Journal of Electronic Dance Music Culture*, 7(2), pp. 5–24.

Westside Gunn x Conway (2016) *Fendi Seats, Griselda Ghost* [Digital Release, Album]. Daupe!

Wilcock, S. (2015) 'The source of magic–rediscovered', *Journal of Performance Magic*, 3(1), pp. 36–56.

Williams, J.A. (2011) 'Historicizing the breakbeat: Hip-Hop's origins and authenticity', *Lied und populäre Kultur/Song and Popular Culture*, 56, pp. 133–167.

Williams, J.A. (2014) 'Theoretical approaches to quotation in hip-hop recordings', *Contemporary Music Review*, 33(2), pp. 188–209.

Yanow, S. (no date) *Conversations with Myself Review, AllMusic.* Available at: https://www.allmusic.com/album/conversations-with-myself-mw0000022550 (Accessed: 4 October 2021).

Zagorski-Thomas, S. (2010) 'The stadium in your bedroom: Functional staging, authenticity and the audience-led aesthetic in record production', *Popular Music*, 29(2), pp. 251–266.

Zagorski-Thomas, S. (2018) 'The spectromorphology of recorded popular music: The shaping of sonic cartoons through record production', in R. Fink, M. Latour, and Z. Wallmark (eds) *The relentless pursuit of tone: Timbre in popular music.* New York: Oxford University Press, pp. 345–365.

Zak III, A.J. (2001) *The poetics of rock: Cutting tracks, making records.* Berkeley: University of California Press.

Zompetti, J.P. and Miller, E.L. (2015) 'After the prestige: A postmodern analysis of Penn and Teller', *Journal of Performance Magic*, 3(1), pp. 3–24.

PART 4

Mixing (records within records)

5

MANUFACTURING PHONOGRAPHIC 'OTHERNESS' FOR SAMPLE-BASED HIP HOP

Charles Mudede (2003) explains that in the context of Hip Hop "a turntable is forced to … make meta-music (music about music) instead of playing previously recorded music", and expands that the sampler is "repurposed to turn one DJ repurposing two turntables into a thousand mini DJs repurposing two thousand virtual, mini turntables". Looking at sample-based record production through such a lens highlights the theoretical complexities inherent in pursuing a comprehensive musicological understanding of the artform, as well as the material implications this poses for its practitioners—particularly those exploring alternatives to copyrighted samples as their source content. As demonstrated in Chapter 3, alongside the numerous creative approaches that sprung out of legal and financial necessity in hip-hop practice since the early 1990s (interpolation, live performance, heavily synthetic subgenres), resorting to sample *construction* brings about its own set of poetic-aesthetic issues.

In reminiscing about his creative reaction to that shifting sample-licensing landscape, Hank Shocklee (2004) made a clear delineation between the sonics that can be acquired from recordings ("using different organic instruments"), as opposed to those that can be acquired "off a record". He also went on to associate the effect of the acquired/sampled sonics both with "the feeling you can get" and the resulting aesthetic ("the sound") of hip-hop outputs produced in an era inevitably defined by these changing practices (Shocklee, 2004). These two considerations will remain key foci in this chapter. The first point highlights beat-makers' preoccupation with phonographic sound as an essential source variable that facilitates the sample-based aesthetic. It is not a stretch to suggest that the second point, with its inferred triangle of *sonics-feeling-output*, refers to the impact the qualities of the source material will have on the beat-maker's sample-based creative *process*. As such, Shocklee's delineation shows that even the descriptor 'sample-based', in the context of hip-hop music, requires further unpacking and, arguably, only tells half the story: that of process, not of the qualities of the source. Therefore, if the previous chapter focused on the performative (and mediated) aspects of beat-making to uncover its—mesmerising—effects, this chapter peels the (mixing) layers off phonographic ephemera to uncover their essence and, in turn, to inform (re)construction.

DOI: 10.4324/9781003027430-11

In their "quest for the real" (Marshall, 2006), the issues sample-*creating*-based practitioners now face become the comparisons their works will inevitably attract against an aesthetic bar set by almost four decades of phonographically sourced sample-based Hip Hop.[1] In other words, the question becomes whether self-created source objects can suffice as effective triggers for sample-based production practices; and what qualities should be infused into these source objects, should they prove inspirational to—rather than simply functional for—the beat-making process. Arguably, there had been less need to discuss the phonographic qualities of a source when the source was *by default* phonographic. But the context framed by these alternative practices necessitates an investigation *into* the source's qualities, as well as the way these interact with sample-based processes.

Thus, a useful way to commence the investigation is through Sewell's (2013) typology of sampling, which systematically categorises sample-based layers in Hip Hop according to their source qualities and function. Although—as we have seen in the introduction—Sewell stops at a structural representation of sample-based layers (rather than an exploration of the mixing mechanics underlying their juxtaposition), her classification does enable an initial unpacking of Krims's (2000, pp. 41–54) "combination of incommensurable musical layers" that contribute to the "hip-hop sublime". Sewell (2013, pp. 26–67) classifies samples into "structural" (main groove), "surface", and "lyric" categories, and these into further subcategories according to their instrumental make-up and organisational function. With regard to structural function, Sewell (2013, pp. 26–67) provides the following classifications:

- "Percussion-only" structural types, which contain "sampled drums [that] are looped throughout the new track";
- "intact" structural types that include "every element from the source material, usually drums and at least one other instrumental line" (these have been referred to as 'foundational' samples previously in this book);
- "non-percussion" structural types which are "very similar to an intact structural sample, except that [they do] not contain sampled drums"; and
- "aggregate" structural types which consist of component "layers … sampled from different sources" or "*different parts of the same source*" (my emphasis).

Conversely, "surface" and "lyric" sample types have a more intermittent or ornamental layering function and can be delineated from each other by their intended lyrical intelligibility; "surface" types can be subcategorised into "momentary", emphatic", and "constituent" types—the latter described as "only a beat or a second long", appearing "only once every measure or two", and "layered against the groove" (Sewell, 2013, pp. 26–67). But how do these interact in the sonic domain?

David Goldberg (2004, p. 129) pinpoints where the missing link may lie; citing Wallace and Costello (1990, p. 85) in 'The scratch is Hip-Hop: Appropriating the phonographic medium', he offers a crucial insight: "Rap/hip-hop has been the first important American pop to use digital recording and mixing techniques in the music's *composition*, its *soul*" (original emphasis). He goes on to attribute the defining characteristic of rap music to "spatial modification" expressed via "exploding kicks", "echoing snares, and the sometimes terrifying sonic manipulations of DJ scratches", mapping the creation (composition) and essence (soul) of beat-making to the interaction of sampling and *mixing* processes (Goldberg, 2004, p. 130). Combined with Mudede's interpretation of sample-based Hip Hop as meta-music, this interaction assumes exponential dimensions for the sample-*creating*-based practitioner.

Not only have the mixing practices of sample-based record production not received sufficient attention, but a reimagined approach that involves the construction of source content *first*, inevitably poses questions about the mixing and manipulation of source objects that themselves require prior recording, mixing, and production actualisation. The pursuit of the newly constructed *phonographic* in a meta context, therefore, necessitates a bidimensional examination of mixing theory as it applies to sample-based Hip Hop from the perspective of both the 'source' and that of (its interaction with) the end output. The creative practice experiments that follow, alongside reflective insights drawn from the project's journal, will illustrate some of these complex phenomena, the analysis drawing out what appears as essential in sample-based poetics via the use of autoethnographic strategies. One of the critical themes that will emerge from the autoethnographic approach will be the notion of phonographic 'otherness'. Grappling with this concept, defining it, and examining its mechanics in the context of sample-based Hip Hop will provide the underlying thread to this chapter.

On 'phonographic otherness'

Hearing otherness

The following section is extracted from a journal entry entitled 'Songwriting for sound', and it illustrates the first of a progressive trajectory of insights that has led to coining the term 'phonographic otherness'. In it, I am reflecting on being immersed in the process of attempting to create *adequate* source material for subsequent sample-based composition:

> It hit me that what I have been doing is, creating music in order to *make sound*. The recent 'songs' made this clear ... I have always felt that the issue was never one of borrowing motifs/phrases that gives sample-based Hip Hop its unique signature; or, I should say it is not solely a musical argument ... This would not explain why beat-makers go for *records*, rather than *recordings*. My pursuit throughout this journey has been to understand the sonic variables that explain this differentiation. My process, it seems, has focused on creating musical *excuses*, so to speak, in order to be able to make mini records—phonographic moments, or ephemera. I have been coming up with riffs, jams, overdubs, even songs, as musical seeds that allow me to then create, capture, and manipulate the sound that carries these musical ideas ... looking back at all the instruments laid down at the end of these long sessions, I see sonifying tools which needed musical ideas—musical context—in order to produce meaningful sounds that could then be captured and made phonographic ... The full immersion into these moments has given the resulting objects ... a musical, stylistic, and sonic coherency that makes them feel as separate entities even when they are/become part of a new beat (original emphasis).

To illustrate this *alterity* between a sampled 'object' and the new beat that uses it, it may be worth analysing, first, an example from discography: Westside Gunn's 'Stefflon Don' from *Supreme Blientele* (2018)—produced by SadhuGold and Hesh—provides a case of highly accentuated differentiation between new and sampled (previously recorded) elements. Westside Gunn's voice carries markers of contemporary recording and production techniques (close-microphone recording, enhanced 'presence' and 'air' in the equalisation, and compression stability) all of which differentiate it clearly from the vocal samples included in the looped phonographic sample. Whether the latter is sampled from vinyl or another format resulting in a lo-fi characteristic (or processed with the intention of sounding old and otherworldly), the combination of tremolo/delayed guitar and haunting vocals that it consists of feel decidedly 'other' to the rap, drums, and sub bass that comprise the new elements. Furthermore, the source sample sounds slowed/pitched down, which adds to the less pronounced top end of its spectral content. The two 'streams' so to speak (old and new), become clear at 0:43, when the sample momentarily cuts out.

From a mix perspective—beyond the clear spectral differences perceptible between the new (present, defined) and sampled elements (featuring less clarity, presence, and high-frequency content)—there are also differences on the *depth* axis of the sonic image (as well as the 'speed' of the sounds in terms of their transient/envelope characteristics). The sample feels rather three-dimensional and infused with notable spatial resonances (particularly on the modulated guitar, but also around the vocals). The whole sonic 'bubble' of the sample—to use a visualisation analogy from Gibson (2008)—is held together by its harmonic distortion, the colouration from the master medium, the recording/mixing signal paths deployed in its making, and any playback/recording devices used during sampling. Little effort seems spent on 'gluing' the samples with the new elements (this seems intentional, part of a lo-fi statement), apart from one heavy-handed but effective strategy: the notable compression applied to the whole beat (new and old elements combined) most likely courtesy of SadhuGold's Roland SP-404 sampler (Mlynar, 2018). This strategy makes the featured sample 'pump', expanding and contracting in terms of volume, reacting to the sub bass and drums, at times drowning the kick drum and, at others, allowing the high-hat to jump out of the combined balance.[2] The effect feels extreme but intentional, paying dues to lo-fi influences (such as RZA's production style and contemporary lo-fi Hip Hop), but also rhythmically and dynamically 'marrying' the two streams together in the end production. The sample is indeed treated as a 'featured' entity within the full beat: dynamically pumped, cut twice, kept separate, kept 'other' whilst, at the same time, *integrated* through the heavy compression approach. The ambience surrounding the sample expands and contracts in tandem, creating a haunting dynamic-spatial effect. The following journal entry provides a personal reflection on the resulting sonic experience:

> This *belonging together* of the elements that comprise the sample, this retainment of the sonic world of the sample whilst featuring it within a new beat, and the simultaneous *celebration* (in terms of production choices) of its 'otherness' whilst integrating it into the new musical context (e.g. chopping, pumping with the beat) is a defining sonic characteristic of sample-based Hip Hop. Sample-based Hip Hop borrows, features, and manipulates not elements, but full masters, expanding and reshaping complete mix 'staging' that has already been committed to a master. As a form of not just music making but also music mixing, sample-based Hip Hop is defined by the sound of the coming together of full mix 'stages' against manipulation possible through sample-based processes. We are actually hearing both new programming and new mixing interacting with previously committed mix stages—so, it is not just the sound of 're-imagined' sequences or phrases, but also the sound of creative ways of integrating phonographic sonic objects (whole 'mix architectures') into meta phonographic processes (original emphasis).

Perhaps, SadhuGold's collaborator, rapper Estee Nack, summarises the effect most succinctly when describing the beat-maker's style as "some old outer space shit" (Mlynar, 2018). In this laconic—if somewhat street—characterisation, the MC zones in on two important conditions for the perception of sonic otherness, as will be examined next: manifestations of time (*old*) and space (*outer space*) featured within the sonic discourse of the sample-based composition.

Defining sonic otherness

I have been using the notion of phonographic 'otherness' to refer to sonic characteristics of source objects in the context of a form of music/making that has been described as meta-music (music about music) (Mudede, 2003). From an autoethnographic perspective, it is important to interpret my use of the term as a sample-*creating*-based practitioner, but also to define otherness more widely. Dictionaries range in their definitions of otherness, from "the quality or *fact* of being different" (Oxford Dictionary, 2019, my emphasis) through to "being or *feeling* different in appearance or character *from what is familiar, expected, or generally accepted*" (Cambridge English Dictionary, 2019, my emphasis). As may be extrapolated just from these two definitions, interpretations of otherness refer to some notion of alterity or difference, but there is no consensus on whether the inferred quality is regarded as absolute or relative. Furthermore, there are multiple understandings of the term in philosophy, psychology, sociology, and anthropology linking otherness to intersubjectivity and social identity, with implications that range from the construction of a self-image, through to attributing otherness "less to the difference of the Other than to the point of view and the discourse of the person who perceives the Other as such" (Staszak, 2009, p. 1). Applying characterisations of otherness to a group, thus, may also be driven by discrimination and so the term has assumed negative connotations in disciplines such as anthropogeography. Staszak (2009, p. 2) provides a helpful delineation, however, stating that "difference belongs to the realm of fact and otherness belongs to the realm of discourse". Ihde (2012, p. 41) adds "that what makes any object 'transcendent,' having genuine otherness, is locatable in this play of presence and absence-in-presence in our perception of things".

For a musicological understanding of otherness, it is useful to turn to Weheliye (2005) who offers a fascinating link between the possibilities offered by the mechanical reproduction of sound (e.g. the phonograph) and notions of (inter)subjectivity as expressed by contemporary Black artists. In *Phonographies: Grooves in sonic afro-modernity*, he demonstrates how fictional characters in modern film/literary narratives:

> ... control and manage the contingencies of sonic otherness by locating it in the sounds of specific subjects ... Music, and sound in general, roots subjects in their environment by making that environment audible, while the immersion that comes with the listening experience is always tied to a space from whence it originates, thereby spatially marking the sound.
>
> *(Weheliye, 2005, pp. 111–112)*

Weheliye here not only demonstrates how the process of mechanically capturing and reproducing human sounds (e.g. music) transfers the energy of a subject onto a localised source, but also illustrates the spatial implications of this sonification. Although Weheliye is primarily concerned with how the sonic reproduction of music expresses the representation of identities negotiating social spaces, it will be interesting to expand on the implications of this idea beyond music consumption/reception/playback and onto music *making*. According to Frith (1996, pp. 123–124), directly engaging with:

> music making and music listening … works *materially* to give people different identities … we're dealing not just with nostalgia for 'traditional sounds', not just with a commitment to 'different' songs, but also with experience of alternative modes of social interaction.
>
> *(original emphasis)*

In his exposition of the turntable as a repurposed or estranged object, Mudede (2003) helpfully explains that: "For Heidegger, a broken object exposes its thingness; for Marx, it exposes its source, the laborer, the one who has transferred his/her body's energy into the substance of the object". The estranged, broken, or repurposed object here is the turntable—Hip Hop's original instrument—transcending from playback tool to music-making instrument, and the source it exposes is the original labourer (the musician/s) whose energy has been materially and physically engraved onto the phonographic record being manipulated. Mudede (2003) illustrates the concept on his blog by depicting hip-hop producer Eric Sermon operating a mixing board, on top of the image of a DJ scratching a record, itself sitting above a picture of Marvin Gaye playing the piano. The illustration could easily be reimagined to feature a beat-maker operating a sampler (with mixing functionality), itself replacing multiple turntablists manipulating/scratching a number of records, which in turn contain recordings (*productions* to be accurate) of live performances (see Figure 5.1). This visualisation helps conceptualise the *meta* levels of sonification involved in sample-based Hip Hop, as well as an illustration of otherness as the sonic alterity of a/multiple subject/s whose essence has been transferred onto material form (the phonographic groove).

It is important to note that Weheliye (2005, pp. 111–112) ties the listening experience to a "space from whence it originates … spatially marking the sound" and that he refers to "contingencies of sonic otherness" in relation to *control*. It would not be a stretch then to reimagine a sample-based producer's (e.g. SadhuGold's) manipulation of a sonic object (for example, a previously released record), not only as an abstract/motivic manipulation of musical material, but as a form of "discourse" (Staszak, 2009, p. 2) or "social interaction" (Frith, 1996, p. 124); in the context of which, the beat-maker exercises control over the material manifestations of recorded subjects' labour. The leap from social spaces to sonic objects is made possible via Weheliye's idea of sound rooting subjects in particular environments (via phonographic playback). The notion of environment, though, can be expanded beyond the spatial to all types of context 'marked' by the phonographic process (geographical location and/or hyperreal space, as well as the era, style, or time communicated by the record). The variables "marking the sound" (Weheliye, 2005, pp. 111–112) become indicators of sonic otherness, a phonographic 'territory' that may resonate both time *and* space (alongside further musicological signifiers). Pickering (2012, pp. 25–26) coins the term "elsewhen" to highlight "the temporal distance brought about by recorded music" noting that: "Musical repeatability means that we are able to hear music from various previous periods and identify them, even on a decade-by-decade basis, by their *characteristic musical sounds*" (my emphasis).[3] It follows, that the sample-*creating*-based practitioner is tasked with

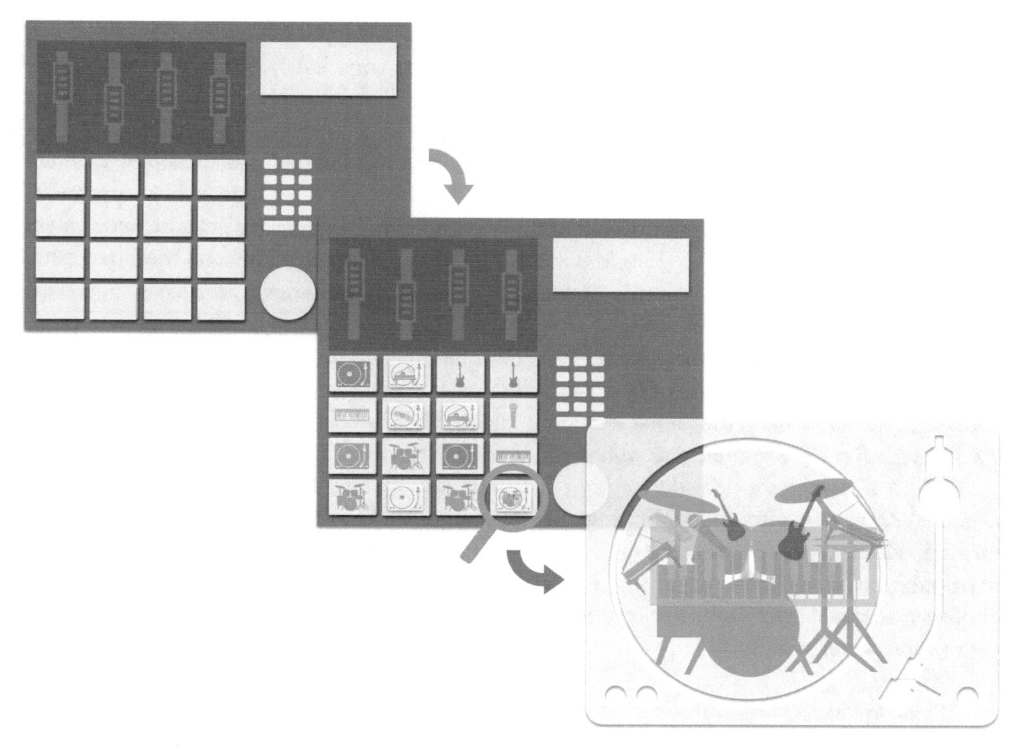

FIGURE 5.1 A schematic representation of a digital sampler (with mixing functionality), enabling the manipulation of multiple record segments, which in turn contain productions of live performances.

the dual objective of not only manipulating (discoursing with / exercising control over) sonic objects that carry identifiable phonographic context, but also with creating and infusing these objects with sufficient sonic 'identity' (character), so that they feature as 'other' against the meta (sample-based) process. But how does this infusion manifest in practice?

Featuring otherness

For the practice-based part of this investigation another original hip-hop production has been conceived ('Call'), built upon two groups of constructed samples. The two-stage process has involved creating and producing the samples as fully committed productions (records) of specific and different stylistic foci (at different times), and without a preconception of what form the ensuing hip-hop production would take. The first of the samples has been sourced out of a multitrack production for a forthcoming release with rock band FET 47. The second sample is a segment taken from an original blues composition for which I have performed and overdubbed all the instrumental layers (drums, electric bass, electric guitar, piano, and harmonica), recorded, mixed, mastered, and produced with particular attention paid to achieving late-1950s Chicago Blues timbral and spatial signatures (in a similar vein to the exposition of Chapter 1). The two source scenarios have been purposefully included, in the first case, to allow access to individual multitrack elements and, in the second, to limit access to the full blues stereo 'master' alone. The intention has been to create an applied context of sources being featured within a new phonographic construct, illustrating consciously featured phonographic contrast as a key aesthetic driver for the envisioned sample-based hip-hop output. Additionally, the different degrees of access (individual multitrack elements vs. full master), allow both the construction of "aggregate" and the use of "intact" structures if one was to deploy Sewell's (2013, pp. 26–67) typological descriptors. As there has been little analysis in literature so far that focuses on the sonic (mixing) aspects of the phenomenon, the creative practice experiment has provided an opportunity to study the mechanics of sample layering beyond their structural functionality (i.e. in terms of the mix architecture). Table 5.1 provides a summarised description of the sample layers, their types, as well as the creative processes (layering and manipulation) that have led to their mix placement in the final hip-hop production. Figures 5.2a and b offer a schematic representation of how the individual layers are 'staged' in the final production. Video 5.1, available from the 'Book > Videos' menu tab at www.stereo-mike.com, showcases the individual sample segments as well as the complete instrumental production.

Table 5.1 provides a neat delineation between musical/abstract processes (surface phenomena) such as re-pitching, chopping, and layering noted in the 'description' column, and mixing/material processes (resulting in staging phenomena) such as the spatial and timbral manipulations detailed in the 'processing' column. It can be summarised that the processing choices associated with both the creation and manipulation of the samples have focused on two overarching strategies:

1. The *narrative/aesthetic*: infusing the samples with characteristic timbral and spatial qualities. For example, the blues 'master' has been created with considerable effort dedicated to reconstructing not just the spatial qualities of late-1950s Chicago blues recordings, but also timbral/tonal signatures reminiscent of the era and representative record-label aesthetics. This has been achieved through the choice of instruments, recording equipment, microphones, and spaces deployed, as well as the emulation of vintage processors (and workflows) used in post-production. On the other hand, the main aggregate (structural) sample made out of band multitrack elements has been sampled through a tape effects emulation pedal and layered with vinyl crackle to construct a non-specific, yet clearly *vintage* record illusion. By way of a tape machine implied as both recording and mixing medium, and vinyl as the final/master format, the recording space shared by the drums and organ, and the matching tracking equipment signatures imprinted on both the drums and electric bass (via Universal Audio hardware compressors), have been accentuated and 'glued' back into a unified sonic experience inferring a shared phonographic time and space.
2. The *pragmatic/architectural*: ensuring the samples work as part of a coherent mix balance and slot in within its overall staging. Much of the filtering, equalisation, and spatial processing decisions have aimed at allowing the juxtaposed samples' full mix/master spectra, stereo images, and depth illusions to fit—in a coherent sense—over each other *and* in combination with the new beat elements (the electric guitar, drum hits, and synthesiser pads).

Returning to the notions of 'elsewhen' and 'elsewhere' as key characteristics of sonic otherness, it is clear to see that the first strategy is responsible for, initially, imbuing the source material with narrative signifiers that tie them to different eras and styles (specific or unidentified), as well as spaces/locations (whether geographical, actual, or hyperreal); followed

TABLE 5.1 A summarised description of the sample layers, their types, as well as the creative processes (layering and manipulation) that have led to their mix placement in the final hip-hop production

Sample/type	*Description*	*Processing*
1: Aggregate main (structural)	Aggregate structure functioning as the main groove (riff/hook) of the new production, constructed out of chopped and pitched-down Hammond C3 organ, acoustic drums, and electric bass performances from the FET 47 multitrack, plus added vinyl crackle. (The syncopated organ part in the verse is replaced by a legato variation for the chorus sections)	Organ sampled through Red Panda Tensor (tape effects emulator) pedal, processed through Akai MPC X amp distortion (emulation), and sent to backwards tape reverb on the sampler; drums sampled through Tensor pedal, processed through VCA-style compression, and sent to a drum room reverb on the sampler; electric bass sampled through Tensor pedal
2: Intact 1 (constituent, surface)	Intact sample taken from the blues production/master and used as a constituent surface sample. (The severe filtering focuses the spectrum on the piano part contained within the intact sample)	Late-1950s Chicago Blues (e.g. Chess Records) inspired recording, mixing, and production process, deploying real recording spaces, a tiled bathroom as echo chamber in post-production, and hardware/software emulations of vintage pre-amps, channels, and outboard processors in both the tracking and mixing phases; end output low- and high-pass filtered, then sent through drum room reverb on the sampler
3. Non-percussion live layer 1 (constituent, surface)	A directly recorded, then pitched up and reversed lead Telecaster guitar performance layer taken from another production; used as a constituent surface sample	Processed through MPC X amp distortion (emulation), and sent to backwards tape reverb and a triplet (dub) delay on the sampler
4. Voice (constituent, surface)	Vocal sample from MPC X onboard library	Sent to backwards tape reverb and a longer non-linear reverb on the sampler
5. Non-percussion live layer 2 (aggregate, structural)	A directly recorded, and then pitched up and reversed rhythm Telecaster guitar performance layer taken from the same production as sample 3 above; used as an additional layer to the main aggregate structure for variation	Processed through MPC X amp distortion (emulation), and sent to the longer non-linear reverb and the triplet (dub) delay on the sampler
6. Intact 2 (aggregate, structural)	Another intact sample taken from the blues production/master, this time used as an additional layer to the aggregate structure	Production master achieved as described in 2 above; then processed through amp distortion (emulation), equalised, and sent to the drum room reverb and triplet (dub) delay on the sampler
7. Intact 3 (aggregate, structural)	As above, equalised and filtered to accentuate the piano part, and used as an additional layer to the aggregate structure	As above, then processed through amp distortion (emulation), equalised, high-pass filtered, and sent to the drum room reverb and triplet (dub) delay on the sampler

Samples 1–4 feature in the main (verse) section of the end production, while the remainder samples (highlighted by shaded cells) are added layers brought in for structural variation and the differentiation of the chorus section.

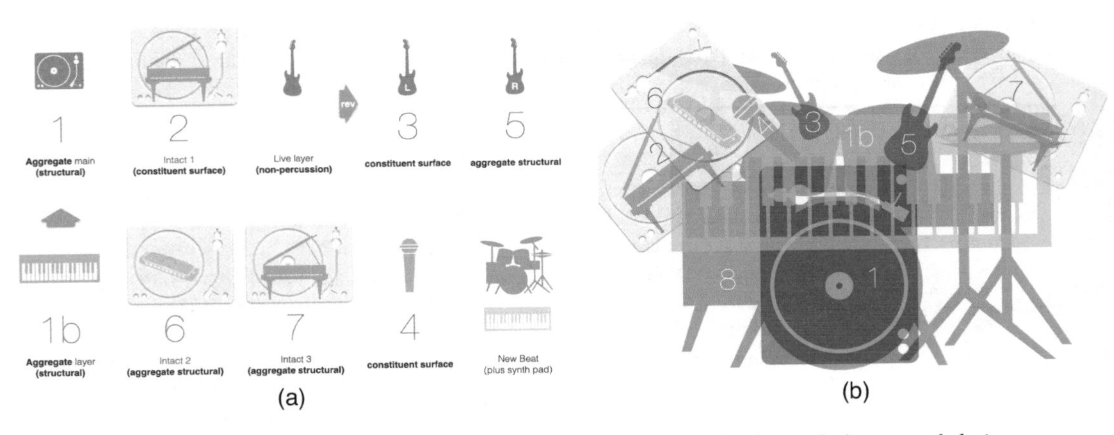

FIGURES 5.2 (a and b) A schematic representation of the individual sample layers and their staging in the final sample-based production (the numbering of the sample representations corresponds to Table 5.1). Note that samples 2, 6, and 7 are represented by a turntable framing the instrument most accentuated by filtering, indicating *intact* structures (mini records within the record).

TABLE 5.2 A typology of sonic characteristics communicating featured 'otherness' in source samples

Sample characteristics defining featured 'otherness'	*Examples*
Limitations in the source's frequency range	vinyl resolution
Recording signal path colourations	microphones used, sound of mixing desk, recording media, outboard equipment
Mixing signal path colourations	sound of mixing desk, recording media used in playback mode, outboard equipment
End format/medium/master sound (colouration, distortion)	master tape, vinyl
Shared captured spaces over recorded elements	recording (live) rooms, echo chambers
Shared spaces applied in the mix/post-production	spatial processors, echo chambers
Playback devices/formats used to record samples	vinyl player, DJ mixer, YouTube/Spotify codecs
Sampling devices/formats used to record, manipulate, and play back samples	phono inputs on sampler, digital extraction codec, virtual software sampler algorithm, filtering, pitch shifting
Surface noise resulting from various mechanical/magnetic production phases	vinyl crackle
Staging architecture is achieved as a result of mixing decisions in three dimensions	stereo width, frequency 'height', (spatial) depth illusion
Mix-buss processing, and colouration when hardware/emulation is used	shared equalisation, dynamic processing, stereo enhancement, sound of outboard
Mastering processing, and colouration when hardware/emulation is used.[a]	shared equalisation, dynamic processing, stereo enhancement, sound of outboard
Purposeful accentuation of source's lo-fi-qualities	quality/resolution reduction, increase/addition of surface noise

[a] Moore (2019, pp. 210–213), indeed, attributes a purposeful "coaxing" of colouration out of recording and mixing equipment by audio engineers he interviews—these devices include microphone preamplifiers, mixing consoles, dynamic range compressors, equalisers, and recording media such as magnetic tape.

by manipulating the sources to negotiate (amplify/intensify or control/limit) these sonifications in the context of sample-based production. The second strategy is concerned with integrating the ensuing sonic contrasts back into a phonographic *whole*, but the primary objective here is 'architectural'—the elements have to be mixed so that the actual frequency spectra, stereo width, and illusory depth of the 'collage' function in a sound-engineering sense.

Although the two strategies are not as clear-cut or always consciously deployed during creative practice, this theoretical delineation helps illustrate the rationale behind the mechanics that communicate aspects of sonic otherness as part of a phonographic context. Table 5.2 provides a typology of the perceivable sample characteristics that define this featured, phonographic otherness.

Further practice-based illustrations

Thus, all of the source content mixed in preparation for the beat-making phase in this project, features—at the very minimum—recording, mixing, and mastering signal-path colourations, as well as staging manifestations, which communicate a temporal and spatial sense of 'belonging' to an/other phonographic context/s. But a number of beats sonically explore more extreme sonifications of otherness, distancing source material even further by means of explicitly stated timbral and spatial manipulations. 'Keim Intro', for example, features an intact sample from previously released work as its foundational layer (the 'Intro' to *The Diary Of Keim Thomas* (Exarchos, 2003)), and a lyric layer from 'Neon Aeon' (from the aforementioned *Gutter Turtle Records* body of work). The former sample has been recorded to (physical) cassette, then played back via the Foldy Makes Sidecar—a modified Walkman with a playback speed-control dial—enabling 'varispeed' (simultaneous pitch and time) manipulation of the recorded audio content. Not only does this process result in striking performative 'warping' and detuning artefacts over the source audio, but it also infuses it with lo-fi tape media characteristics that further distance the source—the intention here being to imbue yet another layer of alterity upon past/own content. The latter (lyric) sample taken from 'Neon Aeon' is recontextualised out of its original mix (but retaining its original sonics/staging), by means of the—more radical—tape effects emulator Tensor pedal (as previously used on the organ sample indicated in Table 5.1), to create an otherworldly, earie, glitch quality upon Jo Lord's vocals. Created out of blending extreme pitched-up manipulations over the pitched-down sample of her original performance, the related pitch and speed ratios 'play' with the thresholds of human psychoacoustics; these range from utterances pertaining to the pitch domain (detuning), to artefacts perceived as delays in the *spatial* domain. Combined with exaggerated manipulations audible at the extremes of the stereo field, the resulting effect contributes to a simultaneous time-, space-, and lateral image-based 'otherworldliness' that resonates with the (re-sequenced) lyrical content.

The cassette warping strategy is redeployed over the foundational content of beats such as 'Keim Outro' (made from similar source content as 'Keim Intro', albeit juxtaposed with vocals from the previously described slam/punk poetry jam in Chapter 3); and 'Gimension', 'Contacts', 'Genitive', and '1998' (all with intact samples taken from old cassette recordings of original content dating back to 1998). Specifically, 'Contacts' uses a surface/lyric layer signifying the first experiment with the sidecar device—the words "stop recording" becoming an ornamental feature over the beat. Conversely, the warped intact piano sample in 'Tense Minor' (made out of the previously discussed americana material) is an artefact produced through manipulation via the tape effects pedal, as is the case with the lyric layer on 'Wishbone', and the pitch-effect articulation of the Rhodes electric piano in 'Billinspired'. 'Balkan Blues' takes advantage of multiple, progressive layers of manipulation over intact samples from the reggae/dub source phase, both through time-based warping via the physical tape-effects pedal and aggressive software distortion within the MPC mixing environment. The foundational loop is layered with elements that are forced to blend in through shared processing characteristics: firstly, a similarly distorted melodica top-line layer (from the same reggae-dub content, but isolated); followed by a wide, stereo instance of a santouri patch performed on a Roli keyboard, with one side of the output manipulated differently to the other through tape effects pedal warping (simultaneously to the playing).[4] The latter strategy is redeployed for the bouzouki-style-patch stereo manipulation on 'Mediterror', over matching—but reverse in their image-processing—harp, strings, and percussion layers taken from a Mediæval Bæbes engineering session.[5] The 'Balkan Blues' santouri chops, including their stereo staging, are re-pitched and chopped again, before being incorporated into the final sections of the 'Mediterror' beat to provide dynamic and textural variation. In these cases, the aggregate structures evolving during the beats unfold a complimentary juggling of processed sonic utterances in the stereo domain, which are further taken advantage of within the MPC's mix environment via the deployment of stereo widening techniques.

In the case of the old cassette recordings, the heavy-handed tape-warping strategies offer a means to impose temporal and spatial distance upon original content that was conceived before or outside the frame of this investigation. It is interesting to consider the implications of media-based staging and sonic manipulation as ways to creatively 'stamp' audio with temporal/timbral and spatial signifiers that transport them to an interim context other than their original one. The rationale of the transformation may be naïve (experimental) or cynical (planned)—its effect, a transportation of the material into a new state of sonic 'limbo', offering reimagined opportunity for the sample-based process. The notion of manipulation as both distancing and transformative effect has been practically explored in a number of further creative scenarios. 'Oneother' and 'Outmosphere' are collaborations with Albin Zak, respectively chopping up his tracks, 'One Another' and 'Your Outer Atmosphere' (from *An Average Day* (Zak, 2002)), after processing them with a software emulation of a 1970s Soviet wire recorder. Described by the designers as a "*magical lofi-tool* and *ghostly echo machine*" (AudioThing, no date, original emphasis), the wire recorder timbral footprint and variable time features (echoes) have enabled an 'altering' of the sonics in Zak's digital masters, which proved fertile to the sample-based "fantasy variations", to re-cite Ihde (2012, p. 108). Furthermore, an old, lengthy cassette recording of free-jazz improvisation between a life-long musical colleague on electric guitar and myself on electric piano—entitled *My Brain Is Bleeding on a Train* (circa early 2000s)—becomes the subject of severe, performative media- and effects-based manipulation. The resulting artefacts range from glitched, lo-fi revenants of electric piano and guitar—retaining some remnants of their harmonic and melodic figures—through to synthesiser-like textures, and completely recontextualised, distorted noise. These become the foundational sample-based layers behind 'Brain Train', 'Train Brain', and 'Freight Train', respectively.

Conclusions

In closing, the characteristics outlined in Table 5.2 are extracted from the aural analysis and creative practice stages of the earlier studies in an attempt to systematise the processes and ensuing signatures that infuse sources with a particular sonic identity. The resulting character differentiates them from new beat-making elements and fuels the sample-based sonic discourse by enabling the interaction of meta-process, and sources perceived as 'other'. This is the aesthetic condition that Schloss (2014, p. 159) refers to when pointing out that "to appreciate the music, a listener must hear both the original interactions and how they have been organized into new relationships with each other". Although each of these characteristics communicates some aspect of sonic otherness, it is important not to think of them as defining variables that explicitly or individually ensure its perception. Instead, collectively, they represent sonic manifestations of 'original interactions' that have taken place as part of a (mini) record-making process: it is the construction of the sources as part of a phonographic vision (a record-making context) that makes them stand out from mere 'recordings' (and, arguably, sample libraries, too), even if instrumental elements/layers end up being used in isolation by way of equalisation, filtering, or through access to multitracks. Zak (2018, p. 304) illustrates this quality best by providing the following disclaimer about record production after the post-war era: "Instead of simply *recording* performances, the idea was to *make records*, with the intent of imbuing the disc with a distinctive personality" (my emphasis).

Of course, the otherness that is communicated by these sonic characteristics works in tandem with musical (harmonic, melodic, rhythmic, stylistic) coherency and structural manifestations (cuts, looping/repetition) that further tie the source utterances together. Moreover, the difference can become accentuated by other bipolarities typically delineating contrast between sample sources and additional beat-making elements, such as: live feel versus programmed quantisation (rhythmic); acoustic and/or electromechanical versus synthesised textures; analogue versus digital colouration (timbral); and spatial decays shared over source elements versus the gated (abruptly and unnaturally truncated) ambient envelopes inherent in single drum hits typically deployed in beat construction (spatial). Finally, the majority of the variables listed in Table 5.2, as well as the signatures enforced by the sample-based production environment (sampler/DAW), may also describe the ways in which the final production of the sample-based artefact integrates the contrasts back into a cohesive end phonographic construct (when the respective techniques are reenlisted as part of the sample-based engineering process).

The sample-based hip-hop aesthetic is the sound of manipulating and recontextualising characteristics (sonic signatures) derived from phonographic ephemera. These characteristics include signal flow colourations and staging phenomena. But if otherness equates perspective rather than just difference, the *meta* process (sample-based composition/production) has got to sonically manifest 'perspective-ness': the sound of discursive workflow, manipulation, a meta-phonographic process interacting with manifestations of—past/other—phonographic processes. In other words, for recontextualisation to function, it has to assume an initial context and, therefore, source samples need to carry markers of having first belonged to a sonic 'elsewhen' and 'elsewhere'. Echoing Schloss, the sample-based artefact sonifies the process of (re)contextualisation—as perspective, as meta-process—within the temporal confines of its structure. However, this sonification does not only manifest in the musical interactions between meta-organisation and original interactions, but also in the mixing (sonic) mechanics that carefully negotiate the dynamics of *contrast* and *integration* through the materiality of textural and spatial manipulation. The autoethnographic lens deployed here has exposed intrinsic aspects of a creative praxis that attempts to construct convincing phonographic 'others' in a sample-based context (making records within records). The examination potentially illustrates how simply making a record is conceptually different to making a record that will feel 'other' within another record, at the same time highlighting the opportunity—and need—to further study the rich sonic phenomena that lie under the surface of contemporary, technologically-interdependent musical forms.

Recommended chapter playlist
(in order of appearance in the text)

'Call'
'Keim Intro'
'Keim Outro'
'Gimension'
'Contacts'
'Genitive'
'1998'
'Tense Minor'
'Wishbone'
'Billinspired'
'Balkan Blues'
'Mediterror'
'Oneother'
'Outmosphere'
'Brain Train'
'Train Brain'
'Freight Train'

Notes

1 This period extends from the mid-1980s to now, should we consider Marley Marl's experiments with affordable samplers around 1984 as the starting point (see, for example: Kajikawa, 2015, pp. 164–165).
2 For a detailed discussion of "lateral dynamics processing" in hip-hop production, see Hodgson (2011).
3 Pickering's *characteristic musical sounds* here echo Zagorski-Thomas's (2014) "sonic signatures".
4 Santouri is the Greek variant of the Persian santur or Indian santoor (a kind of stringed dulcimer played with hammers), and here I am channelling tacit knowledge from my childhood exposure to the instrument to perform it expressively via the modulation possibilities enabled by Roli's tactile Seaboard interface—rubbery keys that respond to movement and pressure in multiple dimensions.
5 The beat's name refers to all foundational elements: the Mediterranean bouzouki, as well as the Mediæval Bæbes' session layers. The engineering session contributed to *A Pocketful Of Posies* (Mediæval Bæbes, 2019), and as with work for Sarabanda and Grupo Lokito, I have offered engineering services in return for sampling permission.

Bibliodiscography

Exarchos, M. (2003) *The Diary of Keim Thomas* [CD, Album]. UK: AMG Records.

Frith, S. (1996) 'Music and identity', in S. Hall and P. Du Gay (eds) *Questions of cultural identity*. Los Angeles, CA: Sage, pp. 108–128.

Gibson, D. (2008) *The art of mixing: A visual guide to recording, engineering and production*. 2nd edn. Boston, MA: Course Technology.
Goldberg, D.A.M. (2004) 'The scratch is hip-hop: Appropriating the phonographic medium', in R. Eglash, J.L. Croissant, G. Di Chiro and R. Fouché (eds) *Appropriating technology: Vernacular science and social power*. Minneapolis: University of Minnesota Press, pp. 107–144.
Hodgson, J. (2011) 'Lateral dynamics processing in experimental Hip Hop: Flying Lotus, Madlib, Oh No, J-Dilla and Prefuse 73', *Journal on the Art of Record Production*, 5.
Ihde, D. (2012) *Experimental phenomenology: Multistabilities*. 2nd edn. Albany: State University of New York.
Kajikawa, L. (2015) *Sounding race in rap songs*. Oakland: University of California Press.
Krims, A. (2000) *Rap music and the poetics of identity*. Cambridge: Cambridge University Press.
Marshall, W. (2006) 'Giving up hip-hop's firstborn: A quest for the real after the death of sampling', *Callaloo*, 29(3), pp. 868–892.
Mediæval Bæbes (2019) *A Pocketful of Posies* [2xCD, Album]. UK: Bellissima.
Mlynar, P. (2018) *Hip-Hop Producer Sadhugold Is Quickly Becoming a Name You Should Know, Bandcamp*. Available at: https://daily.bandcamp.com/features/sadhugold-interview (Accessed: 1 December 2019).
Moore, A. (2019) 'Tracking with processing and coloring as you go', in R. Hepworth-Sawyer, J. Hodgson and M. Marrington (eds) *Producing music*. New York: Routledge (Perspectives on Music Production), pp. 209–226.
Mudede, C. (2003) *The Turntable, CTheory*. Available at: https://journals.uvic.ca/index.php/ctheory/article/view/14561/5407 (Accessed: 2 December 2020).
Otherness|Definition of Otherness by Cambridge English Dictionary (2019) *Cambridge English Dictionary*. Available at: https://dictionary.cambridge.org/dictionary/english/otherness (Accessed: 1 December 2019).
Otherness|Definition of Otherness by Oxford Dictionary (2019) *Lexico Dictionaries|English*. Available at: https://www.lexico.com/definition/otherness (Accessed: 1 December 2019).
Pickering, M. (2012) 'Sonic horizons: Phonograph aesthetics and the experience of time', in E. Keightley (ed.) *Time, media and modernity*. Basingstoke: Palgrave Macmillan, pp. 25–44.
Schloss, J.G. (2014) *Making beats: The art of sample-based Hip-Hop*. Middletown, CT: Wesleyan University Press (Music/culture).
Sewell, A. (2013) *A typology of sampling in hip-hop*. Unpublished PhD thesis. Indiana University.
Shocklee, H. (2004) '"How Copyright Law Changed Hip Hop". Interview with Public Enemy's Chuck D and Hank Shocklee. Interviewed by K. McLeod for Alternet.org', 1 June. Available at: https://www.alternet.org/2004/06/how_copyright_law_changed_hip_hop/ (Accessed: 20 July 2020).
Staszak, J.-F. (2009) 'Other/otherness', in N. Thrift and R. Kitchin (eds) *International encyclopedia of human geography*. Amsterdam: Elsevier, pp. 43–47.
Wallace, D.F. and Costello, M. (1990) *Signifying rappers*. London: Penguin UK.
Weheliye, A.G. (2005) *Phonographies: Grooves in sonic Afro-modernity*. Durham, NC: Duke University Press.
Westside Gunn (2018) *Supreme Blientele* [Digital Release, Album]. Griselda Records.
Wires – Soviet Wire Recorder (no date) *AudioThing*. Available at: https://www.audiothing.net/effects/wires/ (Accessed: 29 September 2021).
Zagorski-Thomas, S. (2014) *The musicology of record production*. Cambridge: Cambridge University Press.
Zak, A. (2002) *An Average Day* [Digital Release, Album]. Records DK.
Zak III, A.J. (2018) 'The death of a laughing hyena: The sound of musical democracy', in R. Fink, M. Latour and Z. Wallmark (eds) *The relentless pursuit of tone: Timbre in popular music*. New York: Oxford University Press, pp. 300–322.

PART 5

(Exponential) Re/mastering

6

'PAST' MASTERS, PRESENT BEATS

At the end of his chapter, 'Considering space in recorded music', Moylan (2012, p. 188) poses the following questions:

> [H]ow do we define the activities and states of spatial qualities as musical materials (concepts) or as ornamental embellishments within the musical texture? How do we calculate their impact on the music, their functions and significance?

His call for further "inquiry ... of how space functions in recorded music" follows the proposition of a methodology and theoretical framework that consider the spatial qualities, perceived distance locations, and lateral imaging of both individual elements and the overall sound of records (Moylan, 2012, p. 187). In response, this chapter examines the implications of the spatial architectures that are constructed within records, in terms of their function as source material in sample-based hip-hop practice. The underlying hypothesis is that—unlike Moylan's pop/rock phonographic examples (e.g. The Beatles and Pink Floyd) that are founded on a track-based approach towards the creation of mix architectures—sample-based Hip Hop depends on the juxtaposition, interaction, and mixing of full *masters*. The approach leads to a form of *exponential sound staging* that sees beat-makers carefully negotiating and reshaping often multiple instances of layered master segments and, I will argue that this phenomenon is a defining aspect of the sample-based sonic aesthetic. As an issue that has not yet received sufficient attention, it complicates existing discourse relating to the notion of staging, necessitating further inquiry. The questions this final part of the project pursues, thus, are:

- How do sample-*creating*-based practitioners construct and merge spatial illusions contained within 'masters' used as source material in hip-hop production?
- What are the dynamics of this interaction? In other words, how do beat-makers negotiate the dimensions of *depth*, *height,* and *width* imbued into masters as part of the creative sample-based process?
- And what is the meaning of these exponential staging strategies for the sonic narratives communicated by the end artefacts?

DOI: 10.4324/9781003027430-13

In order to answer these questions, the chapter resorts to the underlying bricolage methodology that combines literary and aural analysis with autoethnographic interpretations of creative practice. Echoing the strategies of previous chapters, the aim here is to allow for the study, respectively of: literature dealing with the notion of staging; previous hip-hop discography containing relevant case studies; and creative practice functioning as an applied context.

Staging literature and hip-hop sonics

The concept of staging was first introduced by Moylan (1992/2014) with a focus on the spatial implications of mediation possible within a mix. Serge Lacasse (2000) explored it further, investigating the effect of textural and dynamic manipulation specifically on the voice in rock production. Zagorski-Thomas extended the definition to include functional and media-based staging, respectively taking into account "the function to which the recorded output will be put" (Zagorski-Thomas, 2010) and the effect of how "particular forms of mediation associated with audio reproduction media have been used to generate meaning within the production process" (Zagorski-Thomas, 2009). Michael Holland (2013) expanded the concept to include the use of acoustic spaces captured in tracking as a form of staging mediation. And Aaron Liu-Rosenbaum (2012) has been tracing musical and narrative meaning in recording studio aesthetics offering an "expanded notion of staging which applies not only to the voice, but also to instruments".

As staging heavily references a visual metaphor for the representation of sonic phenomena, a number of authors have developed intuitive graphical strategies to illustrate the placement, movement, and manipulation of sonic objects within contemporary music mixes. Popular examples include: Gibson's (2008) conceptualisation of mix layers as sonic objects represented in three dimensions;[1] and Moore and Dockwray's (2008) 'sound-box' illustrations, which add "temporal continuity" to their conceptualisation of a four-dimensional virtual performance space. Moylan (2012, p. 167), however, clarifies that "aligning pitch/frequency with elevation ... is not an element of the actual spatial locations and relationships of sounds, but rather a conceptualization of vertical placement of pitch". Cook (2009) goes beyond metaphor and considers the merits of data-driven visual representation for audio analysis, whilst warning against solely empirical or statistical readings of recordings. His position balances the promise of "a visualization based on objective measurement [that] can act as a prompt to further critical study" with a question of whether "empirical ... approaches [can] really help us understand music as a cultural practice" (Cook, 2009, pp. 236–241). Visual analogy is, therefore, widely deployed to enrich literary theorising on the spatial aspects of recordings and the meaning of staging strategies, but the pursuit of thematic, narrative, or cultural implications favours metaphor over objective data representation (as a bridge between textual reification and sonic manifestations of mixing practice). As will be shown next, conceptual visualisation will form a key means of extending staging theory to cover sample-based phenomena. The strategy will focus on illustrating how the (multi) dimensional space of full masters is (re)staged within hip-hop constructs—a notion that will be referred to as 'sample-staging' in the remainder of this chapter.

The central motivation behind pursuing an extension of staging theory to cover sample-based phenomena is that existing discourse uses, as the basis for the development of analytical frameworks, a binary lens focusing predominantly on two levels: that of the overall sound of a record, and that of individual sources. Moylan (2009) asserts that "[t]hese two levels of perspective or detail are what separate the mastering ... and the mix engineer". But when full phonographic master segments are utilised as building blocks in sample-based composition/production, this function has profound ramifications for the meaning(s) of the practice: the beat-maker additionally assumes a *mastering* perspective, working with the overall sound stages of full masters (record segments), yet *mixing* them as individual elements within the sample-based 'collage'. Beat-making practice, therefore, does not only blur the lines between production and mixing (see, for example: Shelvock, 2017, p. 170) but mastering as well, necessitating a rethinking of sample-based source elements as multidimensional sonic objects. In what follows, I will demonstrate the interrelationship between staging mechanics and the essence of the sample-based aesthetic. The following case studies drawn from discography illuminate such mixing/staging phenomena identified in masters used as samples in hip-hop production.

(Illustrating) Sample-staging in discography

Width, height, and media-based staging

Starting from a sample-staging strategy dealing with a practical conundrum first, the following excerpt from a recent article on low-end stereo placement illustrates how Melba Moore's 'The Flesh Failures (Let the Sunshine In)' (from *Living To Give*, 1970) has been (re) staged in Mos Def's 'Sunshine' (from *The New Danger*, 2004), produced by Kanye West:

> Hip Hop producers ... often face the problem of adding a more powerful bass element to a historic loop containing a bass part ... Kanye West solves this by applying mid/side processing to the sample, thus creating ultra wide stereo with a significant dip in low end frequencies in the middle of the image. Into this he places low bass, often only occupying the sub-bass spectrum ... the careful application of the mid/side processing allows for acceptable mono reproduction.
>
> *(Exarchos and Skinner, 2019, p. 89)*

Figures 6.1a and b respectively portray the sampled record's perceived stage, and the way it has been reshaped within the space of West's beat (and Superstar Dave Dar's mix):

(a) (b)

FIGURES 6.1 (a) A schematic representation of the perceived staging of the chorus in Melba Moore's 'The Flesh Failures (Let The Sunshine In)' and (b) its reshaping in Kanye West's production of Mos Def's 'Sunshine' (the new beat elements enter at the end of the chorus, while a different segment from the original is used for the verses).

Although already notably wide—featuring a "diagonal" (Moore and Dockwray, 2008) image with: piano on the left; organ and orchestral elements on the right; lead vocals, drums, and bass in the middle; and different registers of backing vocals spread both left (for low parts) and right (for high parts)—the 1970s master has been further widened on the lateral axis, but also pitched/sped up. The pitch adjustment results in a frequency shift, pushing the spectrum higher, whilst additional equalisation may have been deployed as part of the mid-side processing. Whether the processing has taken place in the beat-making stage by West, the mixing stage by Dar, or as a combination of both, the resulting sonification is equivalent to a series of (re)mastering artefacts: the weakened middle image and shifted frequency spectra may not have made sense as mastering decisions for a standalone release, but in the context of the new sonic environment they function both in terms of mix architecture and, as will be discussed next, in a narrative sense.

A notable amount of vinyl crackle can be heard on the resulting introductory section of the hip-hop production, which may be the result of a particular combination of record player, stylus, and vinyl record deployed, enhanced by the pitch/equalisation adjustments, or even added in post-production so as to accentuate the vintage qualities of the source. As argued in Chapter 3, sample-based Hip Hop has been founded upon the use of past phonographic sources, and therefore featuring the sonic past—in an audible, exaggerated, or even artificial sense—within its contemporary artefacts has become part and parcel of its aesthetic. Going beyond the functional rationale, thus, it can be argued that the combination of lateral, vertical, and media-based staging for the sampled record has thematic and narrative implications, too. The layers of old elements (sample) and new additions (Mos Def's rap; Kanye West's drum hits and sub bass) are communicated as *distinct streams* via their vintage-contemporary sonic-signature binaries. The striking spatial staging enhances the effect and although it may have been initially conceived of as a pragmatic strategy (creating mix 'space' for the new elements in a lateral and vertical sense), it remains congruent to the sonic interplay of 'past' and 'present'. Moylan (2012, p. 176) asks in relation to image width: "Does the size of the source establish a context or reference for other sources?". This example illustrates that, in a sample-based context, the phonographic object *does* indeed, and it does so in a stylistically defining sense: its *poly*-dimensional (not just spatial, but also media-based) staging utterances establish both a functional (mix-architectural) and narrative (communicative of the sonic past) referential canvas, against which the new elements may be positioned.

The idea of 'sonic narrative' is used here in Liu-Rosenbaum's (2012) sense of the word "where changes in spatial or timbral qualities of an excerpt could conceivably convey a sense of goal-oriented movement". It is also worth noting that this particular version of the song sampled from the 1970 release is difficult to source beyond second-hand vinyl, and not readily accessible from streaming or download services. Therefore, it is safe to assume that what we are hearing on 'Sunshine' is a unique sampling occurrence of particular variables (equipment and vinyl record) that have taken place in West's process. As a result, the type of vinyl noise that is audible and the specific rhythm of its manifestation become unique signifiers of the sampling ephemeron on hand—a processual 'footprint' of sorts. Mark Fisher (2013, pp. 48–49) explains in his article, 'The Metaphysics of crackle: Afrofuturism and hauntology':

> Crackle unsettles the very distinction between surface and depth, between background and foreground ... The surface noise of the sample unsettles the illusion of

> presence in at least two ways: first, temporally, by alerting us to the fact that what we are listening to is a phonographic revenant; and second, ontologically, by introducing the technical frame, the material pre-condition of the recording, on the level of content ... we are witnessing a captured slice of the past irrupting into the present.

Depth/proximity

The above case study demonstrates how the lateral and vertical dimensions of width and height are restaged when a full phonographic master is manipulated in the context of a sample-based composition; and how media-based staging can further 'stamp' and accentuate the narrative ramifications. Of course, relative volume reduction, which often occurs as a result of the source's recontextualisation, presents implications also for the depth or distance perception of the sample's position in the new mix architecture. Alongside the control of ambience/reverberation and high-frequency content, volume manipulation is one of the three essential mediation strategies that engineers deploy to communicate the proximity of a source.[2]

In Chapter 4, Figures 4.2a and b illustrated the effect of perceived depth on the interaction of sampled ('A Theme For L.A.'s Team') and new beat-making elements (Marley Marl's 'Musika' featuring KRS-One). Perhaps the most striking effect in 'Musika' is how the rich construction of a multi-layered depth illusion on the original, becomes a discursive feature in Marl's sample-based juxtaposition. Marl's discreet negotiation of the Philly sound's textural and spatial vintage signatures allows him to juxtapose his contemporary sounds (and KRS-One's rap) against a sonic object that feels like an 'echo' of a past perspective—painting, so to speak, his present (pun intended) boom-bap sonics against a three-dimensional canvas that communicates the past. Of course, the perception of proximity, distance, or depth is a negotiation of sonic perspective on multiple levels—for example, between listener and source, and between source and other sources. "The listener ... can be drawn into becoming part of the 'story' (music) or observing the 'story' (music) from some distance" (Moylan, 2012, p. 173). A sample-based composition, however, can additionally carry *a story within a story*, providing a meta-vantage point so to speak, as it presents the possibility of featuring *a record within a record*. But how does one go about constructing such staging interactions within newly created source material?

(Constructing) Sample-staging in creative practice

The autonomous sonic object

Two practice-based scenarios will be reviewed next, where a sample-based composition has been created out of originally produced source/sampling content. Excerpts from the accompanying reflective journal of the process will be analysed as a means to reflexively build upon developing interpretations of the practice. The first practice-based case study concerns the manipulation of a single instrumental element, demonstrating how phonographic processes related to mix staging and mastering help transcend its perceived quality from a mere 'recording' to a 'record', in the context of a sample-based creative process:

> I came across a grand piano recording I had self-captured about a year ago. I used two Neumann U 87 [microphones] over the sound holes of the piano and a stereo ribbon AEA R88 Mk2 facing the piano lid from a distance, giving me both a solid, clear stereo image of the instrument, as well as a warmer, mellow room tone that I could blend in to change its staging. Reacting to the source, I quickly reached out for a vintage (spring) reverb emulation and applied it only to the close mics. I was aiming for a more distant tone and I also wanted to make the piano more three-dimensional on the Z [depth] axis ... I guess I was making it feel *farther away*, both in terms of physical illusion but also conceptually. I was chasing that phonographic 'otherness', quite consciously attempting to make it feel more mysterious (original emphasis).

In terms of creative intent for the piano source, the rationale and process relayed in the excerpt mirror Reynolds's characterisation of the sample collage as 'ghostly' and the sample in 'Musika' as a distinct, three-dimensional sonic object. It is clear that both the recording techniques and the spatial mixing decisions were aimed at creating a sonic object of notable depth and width. The following reflection demonstrates how the sample was 'distanced' even further through a series of mastering processes and conscious media-based staging choices:

> Synchronising the sampling drum machine to the DAW multitrack playing back the piano tracks, I loaded it up with banks of drum samples and sampled vinyl crackle ... I wanted to distance the piano even further. So, I programmed a combination of vinyl noise samples that made the four-bar piano patterns running in parallel feel like they had been lifted off vinyl. I scanned the 35-minute recording of the piano improvisation for inspiring moments and decided to give the piano mix itself some 'colour' reminiscent of past recording eras ... I applied multitrack tape machine emulations to the individual looping piano subgroups and then ran the full piano mix—including the reverb returns—through a mastering equaliser, a mix-bus compressor, and both master tape recorder and vinyl cutting lathe emulations.

Two essential strategies can be extracted from this process, which aim at infusing the source 'master' with a phonographic footprint and distancing it enough against new elements within the sample-based context: first, the selection and layering of convincing vinyl-crackle patterns and textures placed over the instrumental source—"there is ... no myth without a recording surface which both refers to a (lost) presence and blocks us from attaining it", writes Fisher (2013, p. 49); second, the colouring of the 'master' via the simulation of a vintage-informed mixing and mastering signal flow, reminiscent of "a time when recording technology had developed sufficiently to achieve a kind of sepia effect...". Inevitably, the distancing effect pursued is also related to ideas of perceived authenticity and authority tied to the sample-based aesthetic. Zagorski-Thomas (2009) elaborates: "Playing, sampling and pressing a performance to vinyl as part of the creative process were important statements of authenticity within the Bristol sound of artists"; while for British indie rock in the early to mid-1990s "the notion of authority stems from ... the sound of analogue tape and valve or tube amplifiers ... used to *distance* the sound of Oasis ... from the sound of the 1980s" (my emphasis).

The first of the two tracks showcased in Video 6.1 (available from the 'Book > Videos' menu tab at www.stereo-mike.com) corresponds to the end sample-based artefact ('Glitchando') built from the piano source production, and it sonifies the interaction between the 'constructed' sample and the new beat elements. A noteworthy utterance created by the chopping process performed upon the source master highlights yet another important characteristic: at the fourth bar of every A-section four-bar loop repeat, a reverberant, 'ghostly' texture can be heard, rhythmically interrupting the main piano part on the off-beats. This is the result of a motif performed on the pads of the sampling drum machine, some of which have inadvertently been assigned with soundbites of just reverb decay, as opposed to actual piano notes or chords. The monophonic, legato-style mode enabled on the sampler means that moments of fully staged 'architectures' from the piano 'master' are played as if they were notes on a monophonic synthesiser, each new segment muting the previous one still playing. This performing mode—in combination with other unique programming and swing quantisation affordances facilitated by sampling drum machines—results in striking 'staging rhythms'.[3] These could be described as rhythmical shifts between momentary, or at least short, staging architectures 'frozen in time' on the micro-structural level. Holland (2013) cites Lacasse to describe the effect in a macro-structural sense:

> In Lacasse's terms, the use of multiple reverberant signatures as the track's narrative develops ... are directly related to the piece's structure ... the changes in reverberant character function as an example of diachronic contrast, as the various levels of reverberation are experienced relative to others unfolding within the frame of the recording.

Moylan (2012, p. 177) applies the idea to shifts in lateral imaging, elaborating that "patterns of locations ... and the repetitions and alterations of these patterns can create musical interest just as the patterns of changing pitches, timbres or harmonies". In this sense, *staging rhythms* become a unique musical utterance in sample-based styles, with a narrative-structural function; but the *diachronic contrasts* unfold on a micro scale and within the time domain of the 'loop'. Of course, the effect can take an exponential character when the juxtaposition of momentary 'stages' involves multiple sources, rather than multiple sections from the same source, as the next section will discuss.

The multitrack sonic object

The second practice-based case study illustrates the construction of an original multitrack source for subsequent sampling, highlighting a layered approach to the creation of a number of staging manifestations. The source production in this case has been built by overdubbing acoustic drums, electric bass and guitar, Nord organ, and Fender Rhodes electric piano, followed by the juxtaposition of vocals taken from another source production. A guide beat was also programmed on a sampling drum machine in synchronisation with the developing multitrack, to enable an ongoing evaluation of the evolving 'samples' within a sense of the end context. The following journal excerpt describes how the instrumental performances were recorded with a range of spatial enhancements and timbral shaping gradually committed. As an archiving strategy, the track/file names used during recording disclose the range of processing—serially—applied:

> A [track] name such as 'Tele Wah Stone 63 55 Neve Tape' indicates, for example, a Telecaster guitar, played through a Cry Baby Wah Wah pedal, into an Electro Harmonix Small Stone phaser, and finally a Boss Fender '63 spring reverb pedal. The remainder of the name relates to software emulations [also committed during tracking, such as]: a Fender '55 Tweed Deluxe amplifier, a Neve Preamp, and a Studer A800 multichannel tape recorder … [Performing through] both the pedal reverb being tracked and an AKG BX 20 spring reverb emulation [used only as foldback] inspired the performance, but I could also envision the staging of the guitar in the final mix architecture, whilst making complimentary timbral and musical adjustments … I then reached for my Lakland Jazz bass with the LaBella flats [strings] and played very close to the neck (emulating Aston 'Family Man' Barrett's reggae tone) (Johnson, 2014). To compliment the resulting tone, I run the signal 'hot' through a [real] tube preamp, boosted the low frequencies slightly, and hit an optical tube compressor [circuit] followed by a VCA [hardware compressor] shaving off the peaks … [The end result was] tracked through a Studer tape emulation, effectively mimicking a complete classic signal flow for the referenced era (Leggitt, 2016).

The tracking of the guitar and bass highlight the conscious timbre-shaping decisions committed, on the one hand, ensuring a complimentary tone to sonics gradually being recorded (functional aesthetic) and, on the other, communicating stylistic/era signatures of a non-specific, yet vintage quality (narrative aesthetic). A similar approach was deployed when tracking the keyboards, while the vocal parts were captured with a Shure 520DX 'Green Bullet' microphone slightly saturated through guitar-amplifier and tape-recorder emulations (a typical blues harp recording signal flow). As Zagorski-Thomas (2009) explains: "The other common reason for using media based staging in record production is to evoke the sound of a particular (or more commonly just a vague) historical period". These creative strategies are consistent with an aspect of Williams's (2014, p. 201) intertextual understanding of musical borrowing in Hip Hop: he points out that we may be moving towards a focus on sampling *stylistic topics* instead, where "generic signifiers … become more important than the actual identity of the sample". This argument can extend beyond the musical and the abstract, however, to the materially sonic and concrete, as the following journal excerpt also illustrates:

> Once I found a one-bar [drums] phrase that was sitting well … I looped it around with all mics [channels] active and started mixing it. Auditioning it with and without the beat running in sync, I tried to decide which overheads [mics] I should use (I tracked multiple options, so that I could push the drum aesthetic towards different 'eras') … The drums had been recorded through my choice of hardware preamps with some compression and EQ already committed … [I] run a parallel send of the whole drum mix into a 'pumping' VCA compressor, followed by a passive vintage EQ [both emulations]. The highlighted recorded ambience, enhanced 'air', and tonal glue achieved by the New-York-style parallel layer gave the drums a 'phonographic' quality that was complimentary to the programmed drum hits, providing a sense of 'glue' and achieving that live/non-live fusion

that felt stylistically relevant.[4] Taking the beat out, I was surprised by how few of the drum mics I actually needed for the drum-layering effect to work. I ended up with only the stereo overheads and a little kick support ... The overall mix-bus was going through mastering equalisation, mix-bus compression, and master tape emulations, [so] I had been reviewing and working on the drum mix with the 'hindsight' of auditioning it in this more finalised (end-format) fashion.

The journal excerpt indicates that the drums had been recorded prior to the multitrack subscribing to a 'stylistic topic', using a strategy that deployed multiple microphone choices/techniques, which in turn allowed a degree of sonic-signature-shaping flexibility later in the process. The drum production approach demonstrates particular attention paid to expanding the captured ambient characteristics and testing the interaction between the acoustic footprint and the programmed beat. It can be argued that whatever convincing phonographic sample qualities had been achieved, these were the result of the source operating as a blended, yet distinct sonic 'world' or mix architecture contained underneath the beat. Courtesy of: complimentary staging decisions; shared colourations pertaining to deliberate signal flow choices; a conceptual 'inhabiting' of an aesthetic/era that drove both musical and tonal decisions; and the 'glue' achieved by both tracking and mix-bus processing choices. The mix-bus equalisation, compression, and tape emulation gave the underlying master of the recorded performances a tracked-to-a-particular-recording-medium coherence, which both unify it as a mix of performed elements *and* separate it as a phonographic entity from the—new—beat (elements). Figure 6.2 features a collage of photographs depicting the recording sessions responsible for the production of the constructed 'sample'. The following

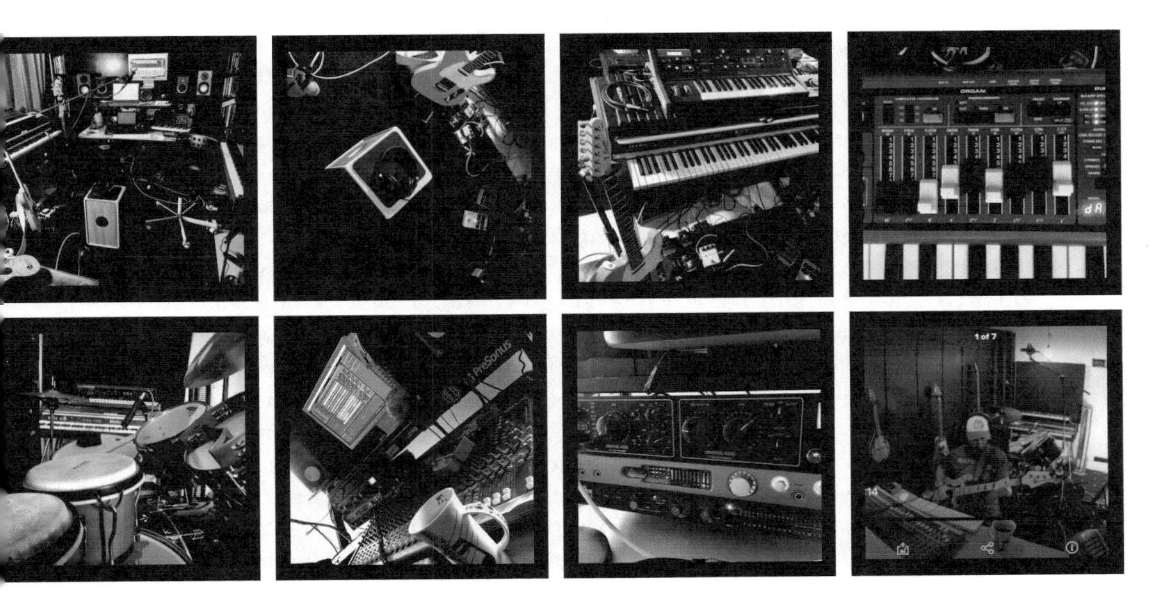

FIGURE 6.2 A collage of photographs from the recording sessions responsible for the production of the constructed multitrack 'sample'.

section will discuss its use for, and incorporation into, the second sample-based composition ('Reggae Rock') under examination.

Exponential staging in sample-creating-based hip-hop practice

In a similar vein to the sampling and chopping processes described for the piano-based production, the beat built out of the multitrack has been constructed by isolating multiple 'staged' moments from the lengthy (approximately 25 minutes) blues-funk 'jam' deconstructed above; pitching/slowing down the samples by −1.63 semitones (about 8.25 beats per minute); rhythmically performing various combinations of the resulting momentary 'masters' using the sampling drum-machine's pads; and further manipulating the segments using the sampler's internal mix functionality. To reenlist Sewell's typological descriptors, the final piece's main A, B, and C sections are created predominantly out of *percussion-only* and *non-percussion* layers, while the breakdown section uses an *aggregate* structure made out of layering multiple component elements sampled from the source multitrack (see Video 6.1); the vocal samples could be described as functioning either as a *surface* or *lyric* type. Of course, access to the component layers is ensured by having created the multitrack production myself, which differs from the possibilities presented by sampling previously-released phonographic material by other artists/producers. The rationale behind working with a range of structural types, here, is driven by the need to test the limitations of access to near-*intact* scenarios, the compositional freedom presented by access to *aggregate* components, but also—importantly—the sonic implications of either approach. As DJ Bobcat (cited in Sewell, 2013, p. 44) explains: "A lot of times when somebody samples a bass and a guitar riff or a horn from the same song, it's because sonically they're the same. They're taking it because they already sound the same". But could this sonic 'sameness' be further unpacked and is it the result of an underlying 'staging harmony' (i.e. a spatial architecture to which all the component layers adhere, even when isolated)? To illustrate the notion of an underlying architecture, Figure 6.3a schematically represents the staging of the multitrack used as the foundation for section A of the sample-based production. Figure 6.3b represents four component layers extracted from different sections of the source multitrack (but retaining their staging placements): a non-percussion layer that includes Rhodes piano, bass, and lead and rhythm guitar (top left); and three percussion-only groupings of cajon-and-bongos (top right), shaker-and-tambourine (bottom left), and drums (bottom right). Note that under the representation of each layer there are opacities overlaid of the missing instruments' positions within the implied mix architecture (providing a kind of *blueprint* for the staging 'harmony').

In order to reinforce the bass part, a matching bass-only layer has also been chopped, equalised, and layered beneath the resulting structure. The perceived effect is of a louder and more prominent bass placement in the main non-percussion layer (the isolated bass layer enables separate equalisation and therefore a complimentary reinforcement of the otherwise harder-to-access bass sonic in the almost intact, non-percussion layer). During the last (eighth) bar of every A-section, two two-beat, non-percussion segments interrupt the main layer on beats one and three to provide a variation and climax (using the sampler's monophonic mode, as in the piano example).[5] Figure 6.3c represents the resulting aggregate structure, as well as the vocal sample juxtaposition, plus the new beat additions (kick drum, snare drum, and high-hat); note the sepia colour added representing the vinyl crackle that has been layered over the aggregate structure. Additionally, the cajon-and-bongos percussion-only

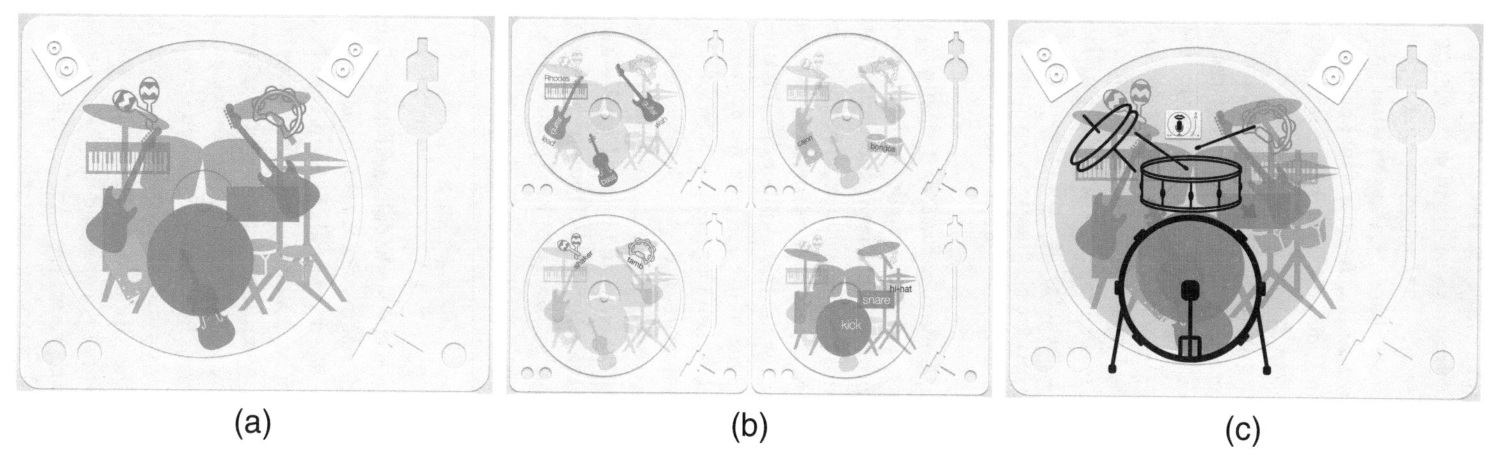

FIGURES 6.3 (a) A schematic representation of the staging of the multitrack used as the foundation for section A of the sample-based production. (b) A schematic representation of the four component layers extracted from different sections of the source multitrack: a non-percussion layer that includes Rhodes piano, bass, and lead and rhythm guitar (top left); and three percussion-only groupings of cajon-and-bongos (top right), shaker-and-tambourine (bottom left), and drums (bottom right). Note that under each layer's representation there are opacities overlaid of the missing instruments' positions within the implied, original mix architecture. (c) A schematic representation of the resulting aggregate structure, plus the new beat additions: the sepia colour added represents the vinyl crackle that has been layered over the aggregate structure; the cajon-and-bongos percussion-only layer has been shifted to the right in terms of lateral imaging, while the arrow pointing down from the new snare and towards the sampled drums' snare represents side-chain compression applied upon the drum layer.

layer has been shifted in terms of lateral imaging, while the arrow pointing down from the new snare towards the sampled drums' snare represents side-chain compression dialled in to reduce the latter's volume with every new snare hit—the strategy aiming at both a balancing and rhythmic interaction between the percussive elements, thus creating complementarity between two initially unrelated samples / drum sonics. A more dynamic visual representation of the staging phenomena is showcased in Video 6.1. Finally, the breakdown section is based on an aggregate structure made exclusively out of component layers, which also include individual Rhodes and lead guitar samples (pitched ten semitones up from the original, which result in an octave interval over the aggregate structure, and twice the tempo).

The aggregate way of working facilitates a refined (re)staging strategy for the component layers. In a sense, the already staged individual, non-percussion, or percussion-only layers are (re)mixed as elements within the sampler's mixing environment. For example, four send effects are deployed (short and long reverb, synced tape delay, and a parallel VCA compressor), which allow sharing/groupings of ambient spaces, mutual rhythmic effects, and common dynamic movement. A number of layers are also individually balanced, equalised/ filtered, and compressed to negotiate the available 'space' more effectively in the resulting sample-based stage.

The intact structural approach—which is more representative of phonographic sampling as with West's and Marl's examples previously analysed—implies a *committed* stage that can only be renegotiated through a form of (re)mastering within the sample-based context. Most pragmatic sample-based creative scenarios fall somewhere between the intact and the aggregate extremes. The added layers (drum hits, etc.) have to interact in a congruent manner in terms of spectra, depth, and width against the three-dimensional frame(s) presented by intact, percussive, or near-intact samples. One of the methods for enhancing this interaction is by integrating side-chain and parallel dynamic processing between the samples, the overall mix, and additional beat elements.[6] Kulkarni (2015, p. 43) muses that "[t]he hip-hop sound always rests on a crucial ambiguity ... that delicious dilemma, that tightrope between looseness/'feel' and machine-like tightness". To echo and expand on Kulkarni—the artform's balance also hangs in the tightrope between contrast communicated by previously constructed ('other') and new elements, and integration (synthesis) achieved through spatial, timbral, rhythmical, and dynamic re-contextualisation.

Further practice

In keeping with the cyclical nature of the project's methodological design, these new understandings drawn out of the interaction of staging theory and practice-based findings, have yet again been applied to further praxis. Tracks 'Arundel', 'Arun Dva', 'Covert Three', 'Spin Eet', and 'Respin'—included in the associated album—exemplify evolving studies of a range of concepts spanning the autonomous-to-multitrack sonic object spectrum discussed here. Specifically, the former three beats make use of cathedral organ recordings deploying multiple-microphone setups to capture the organs in spatially enhanced ways.[7] These recordings have taken place in Arundel and Coventry Cathedrals, as the track names indicate in the phonetic references they contain.[8] Echoing the 'distancing' of the piano and drum sources discussed above, the end beats take advantage of a (dynamic) blend of close and farther microphone positions in the post-production of the organ multitracks. The following 'Beat memo' from the research journal demonstrates how these spatial statements lead to staging rhythms, and how new (synthetic) elements are integrated into the overall spatial illusion:

'Arundel'

Even with a 'single' source there can be a complex context. The Arundel recording is a case in point: one pipe organ, but multiple microphone perspectives (12 channels) positioned as a result of listening to the source in the cathedral, and preempting a mix stage created from the combined perspectives, layered, juxtaposed, even dynamically choreographed.

The interaction and rhythmic interplay of the staged ephemera become a key utterance in the sampled beat: sign-posted by tonal changes manipulated during the performance, varying resonance as a result of the performance dynamics, and spatial call-and-responses from the [ambient] reflections (themselves improvised against, as a play with the space) ... numerous staging rhythms, all made out of them—the wide 0-Coast [synthesiser part] placed in a cathedral [re]verb [in post-production] for matching staging 'epicness'.

Furthermore, the sampled organ layers in 'Arundel' are combined with a percussion-only drum loop—mixed, mastered, and sampled—from a FET 47 band multitrack, to provide the resulting aggregate structures of the beat. Pitched-down and effected speech from yet another (*Songs for Sampling*) source multitrack provides an intermittent lyric layer.[9]

'Arun Dva' starts with organ elements lifted (dug) from the first study, taking advantage of the staging decisions already committed, to explore new musical combinations within a stable (and pre-conceived) staging harmony. The 'play' with(in) this—sparser—stage, here, has led to incorporating further new elements (fretless bass, electric guitar, additional synthesiser pads) into the production, inspired by the sampled drum (indie rock/punk) sonics, and 'glued' via tracking colouration strategies similar to those exposed in the earlier examples (pursuing both complimentary/functional tones and congruent narrative/vintage signifiers). As such, this beat ends up sonically integrating two stylistic streams—the church-organ music, plus the rock-band grouping—underneath the overarching hip-hop sensibility (a third *meta*-stylistic statement), itself communicated by the added drum hits, chopped utterances, and (808) sub bass.

'Spin Eet' and 'Respin' attempt an exposition of two contrasting scenarios. Both beats make use of a nine-minute spinet (small harpsichord) improvised performance, overlaid with mock-operatic vocals;[10] in the first case, sampling the spinet and voice separately (albeit from a combined stage / mix architecture), and in the second case chopping samples of the intact stereo master alongside the individual elements. The source elements have been recorded in the same space and deploying the same microphone, to create a sense of 'duet' interaction (the illusion of singer and accompaniment). The sonic 'gluing' strategies have been further maximised in the mix/master by emulating shared vintage preamplifiers, recording and mastering tape media, console summing, and dynamic and spatial effect processing. In 'Spin Eet', the spinet layer is combined with another (rock) drum source to create an aggregate structure, while the vocal samples mostly assume a surface function. In 'Respin', the beat-making utilises the 'Spin Eet' staging architecture (and elements) as a starting point, but integrates chopping of the added intact (combined spinet and voice) sample as part of the multi-layered, reimagined sequences. As such, this beat/study blends notions of remixing (elements) and more classic beat-making (stereo/master-chopping) approaches, benefitting from the underlying staging harmony, timbral consonance, or implied architecture. In the journal's 'Beat memos' I remark:

> Consider the idea of progressive staging: jamming and developing from previous staging conceptions; the way I use a study to then build a tune, or an alternative or next tune. This is so important, that once a conducive stage is created, *riffing is a jam with the stage ...* Sometimes the mix comes before the composition... (my emphasis).

'Covert Three' takes the notion of exponential staging to the cube, by blending pre-staged spinet and voice layers taken from the two previous studies, with the multi-miked Coventry organ recordings, as well as alt-rock elements such as the distorted drum loops and pitched-down spoken/lyric layer audible. The aggregate structure is accentuated by church bells recorded as Foley (and inspired by the cathedral narrative associations), and the beat is completed by single drum hits supported by 808 sub-bass notes. The multiple elements are 'held' together by added, shared ambient illusions (spatial processing), and a timbral-dynamic gluing strategy that echoes the whole album endeavour: 'pushing' the MPC's converters to the edge (to benefit from second-order harmonics via judicious clipping); summing on an analogue mixer, incorporating parallel harmonic and compression strategies; and carefully gain-staging through a number of analogue transformers in the master signal path, which includes a 'colourful' mix-bus compressor lightly 'pumping' the end program material.

Conclusions

This final chapter has illustrated staging mechanics in sample-based hip-hop phenomena across a spectrum of creative contexts: from phonographic sampling utilising full, previously-released master segments; through to sample-*creating*-based practices that—via extended access to multitrack elements—facilitate a multi-layered approach to the shaping, control, and manipulation of the source's staging dimensions. At the heart of the process, lies a sonic object that carries an extended mix architecture, with the potential to not only provide raw sonic content for this form of material composition (sample-based Hip Hop); but also a poly-dimensional referential canvas that can communicate narrative notions, such as representations of the past, diachronic contrasts, and striking genre-defining utterances such as syncopated *staging rhythms*. These perceptual effects depend on the construction of convincing spatial and media-based staging artefacts for the hip-hop practitioner creating their own source material; and these, in turn, translate to phonographic signatures contributing to an authentic sample-based footprint. It can be deduced that the essential aesthetic of sample-based music forms—and the key differentiation between a generic sonic element and an actual *sample* at the heart of their processes—can be traced in this interaction with staged sonic objects carrying *markers of phonographic process*. It is a manifestation of a phonographic poetics interacting with previously (even if very recently) committed phonographic poetics. This kind of layering can, therefore, become exponential, and sample-based music forms deal not with mixing elements, but with mixing and manipulating full 'masters'—with a playful 'polystagiality' residing at the heart (mechanics) of the creative phenomena.

Recommended chapter playlist
(in order of appearance in the text)

'Glitchando'
'Reggae Rock'
'Arundel'
'Arun Dva'
'Covert Three'
'Spin Eet'
'Respin'

Notes

1 Gibson deploys a vertical/height axis for pitch/frequency, a horizontal/width axis for lateral position, and a depth axis for distance location.
2 Moore and Dockwray (2008, p. 219) often refer to this dimension as "prominence".
3 In this case, an Akai MPC X has been deployed again.
4 This is a reference to a technique characteristic, initially, of New York mix engineers, Owsinski (1999, p. 52) calls the "New York Compression Trick": it aims at enhancing the rhythm section in a mix, by blending a heavily compressed and additively equalised version of—typically—the drum elements, with a more conservatively processed version of the tracked original.
5 The segments are equivalent to the main non-percussion layer in terms of elements included, but with added organ parts—the latter of which features a glissando.
6 This is a strategy indeed championed by this production: a compressor inserted on the overall sampler mix is triggered by the new kick drum sample, while multiple elements are routed to the sampler's parallel VCA-style compression bus.
7 The recordings have been engineered in collaboration with Professor Justin Paterson.
8 In a manner similar to the playful naming schemes we witnessed J Dilla and Madlib deploy in Chapter 4: 'Dva' translates to 'two' in Croatian (one of my two native backgrounds), and it signifies the second study created out of the same Arundel Cathedral recordings; 'Covert Three' is a phonetic play on Coventry, particularly as the organ staging here assumes a background function in the staging architecture of the end beat, whilst also signifying the third study of multi-miked organ 'stages' integrated into the beat-making practice.
9 The narration is a recording of an article I wrote in response to Greek rapper Killah P's murder by a Golden Dawn member; it was published in Sonik magazine's special issue against fascism (Stereo Mike, 2013). As such, 'Arundel', with its spiritual undertones and spatial sense of awe, is imagined as a sonic dedication, and a requiem, for my late peer.
10 The composition/improvisation very loosely channels references from Rammstein, Stockhausen, Mozart's *Don Giovanni*, and favourite Russian baritone Dmitri Hvorostovsky.

Bibliodiscography

Cook, N. (2009) 'Methods for analysing recordings', in N. Cook et al. (eds) *The Cambridge companion to recorded music*. Cambridge: Cambridge University Press, pp. 221–245.

Exarchos, M. and Skinner, G. (2019) 'Bass-the wider frontier: Low-end stereo placement for headphone listening', in J.-O. Gullö (ed.) *Proceedings of the 12th Art of Record Production Conference - Mono: Stereo: Multi*. Stockholm: Royal College of Music (KMH) & Art of Record Production, Us-AB, pp. 87–104.

Fisher, M. (2013) 'The metaphysics of crackle: Afrofuturism and hauntology', *Dancecult: Journal of Electronic Dance Music Culture*, 5(2), pp. 42–55.

Gibson, D. (2008) *The art of mixing: A visual guide to recording, engineering and production*. 2nd edn. Boston, MA: Course Technology.

Holland, M. (2013) 'Rock production and staging in non-studio spaces: Presentations of space in Left Or Right's 'Buzzy'', *Journal on the Art of Record Production*, 8.

Johnson, K. (2014) 'Legend: An Interview with Aston "Family Man" Barrett', *No Treble*, 26 June. Available at: https://www.notreble.com/buzz/2014/06/26/legend-an-interview-with-aston-family-man-barrett/ (Accessed: 1 September 2021).

Kulkarni, N. (2015) *The periodic table of Hip Hop*. London: Random House.

Lacasse, S. (2000) *'Listen to my voice': The evocative power of vocal staging in recorded rock music and other forms of vocal expression*. Unpublished PhD thesis. University of Liverpool.

Leggitt, B. (2016) 'Rhythm Guitar and Other Reggae Secrets', 30 December. Available at: https://planetbotch.blogspot.com/2016/12/rhythm-guitar-and-other-reggae-secrets.html (Accessed: 5 January 2021).

Liu-Rosenbaum, A. (2012) 'The meaning in the mix: Tracing a sonic narrative in 'When The Levee Breaks'', *Journal on the Art of Record Production*, 7.

Moore, A.F. and Dockwray, R. (2008) 'The establishment of the virtual performance space in rock', *Twentieth-Century Music*, 5(2), pp. 219–241.

Moore, M. (1970) *Living To Give* [Vinyl LP]. US: Mercury.

Mos Def (2004) *The New Danger* [CD, Album]. UK: Geffen Records, Island Records Group.

Moylan, W. (2009) 'Considering space in music', *Journal on the Art of Record Production*, 4.

Moylan, W. (2012) 'Considering space in recorded music', in S. Frith and S. Zagorski-Thomas (eds) *The art of record production: An introductory reader for a new academic field*. Surrey: Ashgate, pp. 163–188.

Moylan, W. (2014) *Understanding and crafting the mix: The art of recording*. 3rd edn. Oxon: CRC Press.

Owsinksi, B. (1999) *The mixing engineer's handbook*. Vallejo, CA: MixBooks.

Sewell, A. (2013) *A typology of sampling in hip-hop*. Unpublished PhD thesis. Indiana University.

Shelvock, M. (2017) 'Groove and the grid: Mixing contemporary Hip Hop', in R. Hepworth-Sawyer and J. Hodgson (eds) *Mixing music*. New York: Routledge (Perspectives on Music Production), pp. 190–207.

Stereo Mike (2013) 'Ο φασισμός γύρω και μέσα μας: Οι Έλληνες καλλιτέχνες μιλάνε αποκλειστικά στο Sonik', *Sonik*, p. 27.

Williams, J.A. (2014) *Rhymin' and stealin': Musical borrowing in Hip-Hop*. Ann Arbor: The University of Michigan Press.

Zagorski-Thomas, S. (2009) 'The medium in the message: Phonographic staging techniques that utilize the sonic characteristics of reproduction media', *Journal on the Art of Record Production*, 4.

Zagorski-Thomas, S. (2010) 'The stadium in your bedroom: Functional staging, authenticity and the audience-led aesthetic in record production', *Popular Music*, 29(2), pp. 251–266.

OUTRODUCTION

This book has taken a gradual 'probing' approach to the theoretical/analytical deconstruction and, then, iterative/practical (re)construction of the sonic phenomena that characterise (useable) samples at the heart of the beat-making process. Progressively, the approach has revealed: (i) the sonic/material dimensions behind the inter-stylistic dynamics of sample-based music making; (ii) the effect of sampling technologies on sample-based stylisations; (iii) the importance and manifestation of the sonic past in Hip Hop via the conscious integration of vintage/phonographic signatures in its creative modus operandi; (iv) the perceptual effects and creative potential the interaction between beat-making and multi-layered phonographic objects has on listeners and makers alike; (v) the spatio-temporal, narrative, and thematic implications the 'staging' placement and dynamic mediation of musical elements has for the end constructs/beats; and (vi) the poly-dimensional extent staging practices assume in sample-based hip-hop praxis, thus offering an expansion of staging theory beyond multitracking practices in contemporary record production. By responding to a practical conundrum in current beat-making practice, the investigation has come face-to-face with what constitutes a source sample as 'phonographic' in its essence; and what the sonic dimensions of raw material are such that they, not only contribute to a sample-based aesthetic, but facilitate effective beat-making praxis.

Moreover, the project has contextualised the duality/oscillation between retrospective romanticism (analogue nostalgia) and forward-thinking creative intentionality (digital futurism). These forces are expressed in the consciously (re)constructed vintage sonics and exponential/playful sample-based 'poly-stagiality' as part of a metamodern state of creative affairs. The beat-making praxis materialises these evolving conceptualisations by presenting, in the associated album, a spectrum of end productions that range from contemporary Boom Bap (such as, for example, 'Como Mi Ritmo'); to more experimental sample-based stylisations, echoing what Hodgson (2011) describes as "experimental Hip Hop", or D'Errico (2015) refers to in 'Off the grid: Instrumental Hip-Hop and experimentation after the Golden Age' (for example, in beats like 'Take 3' or 'Train Brain'). Like Hodgson, who focuses on specific processing practices as contributing factors towards

DOI: 10.4324/9781003027430-14

experimental hybridisation, the latter content here is a direct result not only of conceptual 'studies', but also of more extreme mediation practices and spatial processing enforced upon *isolated* multitrack elements—in a process akin to remixing. The bridging with sample-based hip-hop signatures, however, becomes possible (and audible), by ensuring these individual layers carry sonic manifestations of their underlying timbral and spatial staging blueprints in the new constructs, crafted as part of their original(ly imagined) 'phonographic' context.

As such, the project has deconstructed the elements of the sonic domain that constitute the 'mechanical' dimensions of borrowing in sample-based Hip Hop, supporting the initial hypothesis for its significance in beat-making poetics, and facilitating a modus operandi that empowers an aware p/re-construction of raw materials towards it. Additionally, the investigation has expanded upon the original premise, by revealing rich spatial dimensions negotiated in the interplay between beat-making practices and the perceived sound stage of source 'phonographic' objects, reaching beyond vintage surface/textural signifiers. Thus, the notion of 'making records within records' assumes the power of a sonic manifesto that enables the 'exponential' in phonographic poetics, and allows for a (re)imagining of what future meta(modern)-music may sound like. The project's contributions can therefore be summarised in the following statements:

- The expansion of staging theory to cater for sample-based music production practices;
- A phenomenological unpacking of the sonic/material dimensions in sample-based musicking (deploying mixing/engineering expertise);
- Providing a nuanced account of sample-*creating*-based hip-hop praxis;
- Extending the praxis via intentional poly-stagiality through expanded 'depth'/spatial sculpting and timbral mediation in the sonic domain; and
- The proposition of a multi-method framework synthesising a range of strategies that may be transferable to self-study in other arts-based contexts.

Future directions

The emergent focus on the spatial dimensions of staging in a sample-based context, and the practical implications of the concept for sample-*creating*-based beat-making, pave the way for promising future directions in both research and practice. In the current climate of spatial audio enthusiasm, experiments with 3D and/or binaural sonic objects make sense for a form of beat-making that explores not just the illusion of poly-dimensionality packed into mono or stereo 'images', but potentially creative 'play' with actual immersive 'stages' folded into easily managed file formats. Collaborator Jo Lord's (2022) democratic approach to spatial audio developed in 'Redefining the spatial stage: Non-front-orientated approaches to periphonic sound staging for binaural reproduction' presents an attractive opportunity for beat-makers who may opt for access to immersive sonic/phonographic objects via drum machines or software alternatives; and the notion of 'binaural Boom Bap' has already been enacted as part of a collaborative experiment resulting in track 'Flipped' in the associated album (note: the expanded binaural staging will only be perceptible via headphones).

From a practice-based, real-world perspective, the instrumental productions included here have been conceptualised as part of a dual function: as self-sufficient instrumentals, where the vocal 'chops' simply act as surface sounds; but also as beats with lyrical connotations that may inspire future lyric-writing and rapping. Finally, the pedagogical implications of the research have been explored in Exarchos (2018), extrapolating on the collaborative potential existing between performing, sound engineering, and electronic music production disciplines—a notion/model that is transferable beyond the academy.

However, it is important to acknowledge that an essential catalyst in this research (and its practical actualisation) has been the deployment of substantial, pre-existing sound-engineering knowledge, which may be an elusive factor for young beat-makers. Therefore, it is fruitful to consider DIY routes towards the construction of 'phonographic' context in source content, which may be more readily applicable by the beat-making community. On a related note, beats such as 'Arun Dva' (deconstructed in Chapter 6) and 'Rebluezin' (deconstructed in Chapter 2) have demonstrated how new instrumental layers may be 'glued' to sampled source constructs via tracking colouration strategies that pursue congruent sonic signifiers. A promising future direction therefore lies in systematically examining staging intentionalities at the tracking (recording) stage of individual elements, providing a faster route towards 'phonographic' cohesiveness between sampled layers for the contemporary (and future) beat-maker. These may range from era-invoking signal-flow paths (for example, systematically applied pre-amplifiers of recognisable mixing desks at the input stage), to shared spatial resonances (carefully gauged ambient traces of recurring real acoustic spaces or spatial processors captured with every instrument). This approach contains the potential to engage the 'blueprint' of a staging harmony before a multitrack object is even (needed to be) constructed: "Sometimes the mix comes before the composition" (preempted in Chapter 6).

In closing, a reflexive listening analysis of the whole body of work created via self-sampling practices such as those accounted here, may reveal a potent interrelationship between the producer's evolving sonic footprint and recurring source-crafting tendencies (hermeneutically revealing underlying expressive patterns).[1] Albeit benefitting from a certain amount of temporal distance, future analytical work could focus on such relationships, sonically interpreting the implications of self-sampling strategies on evolving stylisation; and examining—through a recording analysis prism—Beer and Sandywell's (2005, p. 119) claim that: "genre … is also a practitioner's term invoked in the recognition, consumption, and production of musical performances". The superpower of a unique and personal production signature—let alone a creative reaction to a licensing headlock—may lie in the generation of both source content *and* its manipulation in a sample-based musicking context.

Recommended chapter playlist
(in order of appearance in the text)

'Como Mi Ritmo'
'Take 3'
'Train Brain'
'Flipped'
'Arun Dva'
'Rebluezin'
'Aldenham'

Note

1 Beats such as 'Aldenham', for example, reveal a 'chanting' signature frequently recurring in the accompanying album: inadvertently, the result of depending on, and then distancing, my own voice as a source element. In discussion and analysis with the project's musical collaborators, a possible interpretation offered has been that this signature may be the result of notable exposure to Byzantine chanting in Greek churches during childhood: perhaps a subconscious influence expressed via the amalgamation of sample-based manipulation and 'digging' for *otherness* in echoes of my own sonic past.

Bibliodiscography

Beer, D. and Sandywell, B. (2005) 'Stylistic morphing: Notes on the digitisation of contemporary music culture', *Convergence: The International Journal of Research into New Media Technologies*, 11(4), pp. 106–121.

D'Errico, M. (2015) 'Off the grid: Instrumental Hip-Hop and experimentation after the Golden Age', in J.A. Williams (ed.) *The Cambridge companion to Hip-Hop*. Cambridge: Cambridge University Press, pp. 280–291.

Exarchos, M. (2018) 'Hip-Hop pedagogy as production practice: Reverse-engineering the sample-based aesthetic', *Journal of Popular Music Education*, 2(1&2), pp. 45–63.

Hodgson, J. (2011) 'Lateral dynamics processing in experimental Hip Hop: Flying Lotus, Madlib, Oh No, J-Dilla and Prefuse 73', *Journal on the Art of Record Production*, 5.

Lord, J. (2022) 'Redefining the spatial stage: Non-front-orientated approaches to periphonic sound staging for binaural reproduction', in J. Paterson and L. Hyunkook (eds) *3D audio*. London: Routledge (Perspectives on Music Production), pp. 256–273.

END CREDITS

Beats *(in alphabetical order)*	*Written by*	*Instruments sampled*	*Performed by*
'1 By 2'	M. Exarchos, P. Thompson, R. Toulson	Electric bass, synthesisers (Korg MS20, Roli), upright piano	M. Exarchos
		Drums	P. Thompson
'1960s 2'	M. Exarchos	Voice	A. Caldecott
		Drums, electric bass, Nord Farfisa organ, percussion	M. Exarchos
'1998'	M. Exarchos	Yamaha W7 synthesiser	M. Exarchos
'Aldenham'	M. Exarchos	Drums, electric bass, Rhodes electric piano, voice	M. Exarchos
'Alderoots'	M. Exarchos	Drums, electric bass, Nord harpsichord, Rhodes electric piano, voice	M. Exarchos
'Altcore'	M. Exarchos	Drums, electric and acoustic guitars, electric bass, Korg MS20 synthesiser, voice	M. Exarchos
'Americanotha'	M. Exarchos	Electric bass, electric guitar, Nord Wurlitzer electric piano, percussion, upright piano, voice	M. Exarchos
		Drums	P. Thompson
'Arun Dva'	M. Exarchos	Electric guitar, fretless electric bass, Pipe organ, synthesisers (Korg MS20, Make Noise 0-Coast, Roli), voice	M. Exarchos
'Arundel'	M. Exarchos	Pipe organ, synthesisers (Korg MS20, Make Noise 0-Coast, Roli), voice	M. Exarchos
'Balkan Blues'	M. Exarchos	Fretless electric bass, melodica, percussion, Roli synthesiser	M. Exarchos
		Drums	P. Thompson
'Become'	M. Exarchos	Drums, electric bass, electric guitar, Korg MS20 synthesiser, upright piano, voice	M. Exarchos
'Billinspired'	M. Exarchos, A. Hector-Watkins	Electric bass, electric guitar, melodica, synthesisers (Moog, Roli), percussion, Rhodes electric piano	M. Exarchos
		Voice	A. Hector-Watkins
		Drums	P. Thompson

(*Continued*)

Beats *(in alphabetical order)*	*Written by*	*Instruments sampled*	*Performed by*
'Boom Bag'	M. Exarchos, D. Pratt	Harpsichord, percussion, Rhodes electric piano	M. Exarchos
		Acoustic guitar, baglama, electric bass, voice	D. Pratt
'Born To Death'	M. Exarchos, J. Lord	Accordion, cigar box guitar, fretless electric bass, kalimba, mandolin, ukulele, upright piano, xylophone	M. Exarchos
		Percussion, voice	J. Lord
'Boxing'	A. Caldecott, M. Exarchos	Voice	A. Caldecott
		Electric bass, electric guitar, upright piano, Roli synthesiser	M. Exarchos
'Brain Train'	M. Exarchos, A. Tsoukatos	Synthesisers (Moog, Roli), Yamaha electric piano	M. Exarchos
		Electric guitar	A. Tsoukatos
'Call'	M. Exarchos, P. Thompson, R. Toulson	Electric bass, electric guitar, Hammond C3 organ, harmonica, upright piano	M. Exarchos
		Drums	M. Exarchos, P. Thompson
'City'	M. Exarchos, P. Thompson	Acoustic guitar, electric bass, electric guitar, Korg MS20 synthesiser, Nord Chamberlin, Nord Hammond organ, Nord Mellotron, percussion, Rhodes electric piano	M. Exarchos
		Drums	P. Thompson
'Como Mi Ritmo'	M. Exarchos	Electric bass, lead voice	E. Caicedo Alegria
		Congas, bongo, campana	B. Bland
		Timbales	T. Butler
		Grand piano	S. McGuinness
		Güiro	D. Pattman
		Backing vocals	E. Solis
'Contacts'	M. Exarchos	Synthesisers (Make Noise 0-Coast, Korg MS20, Yamaha W7), voice	M. Exarchos
'Covert'	M. Exarchos	(Rijeka) Foley, pipe organ, voice	M. Exarchos
		Drums	P. Thompson
'Covert Three'	M. Exarchos	(Rijeka) Foley, harpsichord, pipe organ, voice	M. Exarchos
		Drums	P. Thompson
'Cycles'	M. Exarchos	Drums, electric bass, Rhodes electric piano, harpsichord, upright piano	M. Exarchos
'Dragga Five'	M. Exarchos	Drums, electric guitar, fretless electric bass, melodica, percussion	M. Exarchos
'Flipped'	M. Exarchos, J. Lord	Electric bass, electric guitar	M. Exarchos
		Voice	J. Lord
'Freight Train'	M. Exarchos, A. Tsoukatos	Yamaha electric piano	M. Exarchos
		Electric guitar	A. Tsoukatos

'Genitive'	A. Caldecott, M. Exarchos	Voice	A. Caldecott
		Yamaha W7 synthesiser	M. Exarchos
'Gimension'	M. Exarchos	Double bass, voice	Anon.
		Yamaha W7 synthesiser	M. Exarchos
'Glitchando'	M. Exarchos	Grand piano, synthesisers (Korg MS20, Moog, Roli)	M. Exarchos
'Good Ole Betty'	M. Exarchos	Electric bass, electric guitars, upright piano Roli synthesiser, voice	M. Exarchos
		Drums	P. Thompson
'It Meters'	M. Exarchos	Drums, electric bass, electric guitar, harmonica, percussion, Rhodes electric piano, voice	M. Exarchos
'Kaishaku'	M. Exarchos, J. Lord	Backing vocals, electric bass, electric guitar, Roli synthesiser	M. Exarchos
		Voice	J. Lord
'Kalimbap'	M. Exarchos	Harpsichord, kalimba, voice	M. Exarchos
'Keim Intro'	M. Exarchos, J. Lord	Gamelan, grand piano, Rhodes electric piano, Roland JD800 synthesiser	M. Exarchos
		Voice	J. Lord
		Drums	P. Thompson
'Keim Outro'	A. Caldecott, M. Exarchos, R. Harbron	Voice	A. Caldecott
		Gamelan, grand piano, Rhodes electric piano, Roland JD800 synthesiser, upright piano	M. Exarchos
		Concertina	R. Harbron
		Drums	P. Thompson
'King's Funk'	M. Exarchos	Drums, electric bass, electric guitar, Rhodes electric piano, percussion, voice	M. Exarchos
'La Noche (AMHB)'	M. Exarchos, J. Martinez, S. McGuinness	Backing vocals, electric bass	E. Caicedo Alegria
		Congas, bongo, campana	B. Bland
		Timbales	T. Butler
		Roli synthesiser	M. Exarchos
		Tenor saxophone	C. Hirst
		Electric guitar	K. 'Burkina Faso' Kasongo
		Lead vocals	E. Makuta
		Grand piano	S. McGuinness
		Trombone	V. Msimang
		Güiro	D. Pattman
		Backing vocals	E. Solis
		Trumpet	D. Wilhelm
'Left Right'	M. Exarchos	Drums, electric guitar, fretless electric bass, percussion, Rhodes electric piano, upright piano	M. Exarchos
'Mediterror'	M. Exarchos	Orchestral harp	C. Beer
		Violin	K. Blake
		Synthesisers (Moog, Roli)	M. Exarchos
		Drums	P. Thompson
		Metallic percussion	B. Woollacott

(*Continued*)

Beats *(in alphabetical order)*	*Written by*	*Instruments sampled*	*Performed by*
'Mo' Town'	M. Exarchos, J. Martinez, S. McGuinness	Drums, grand piano, percussion, Rhodes electric piano, Roli synthesiser, voice	M. Exarchos
		Tenor saxophone	C. Hirst
		Trombone	V. Msimang
		Trumpet	D. Wilhelm
'Negrito Blues'	M. Exarchos	Electric bass, electric guitar, Rhodes electric piano	M. Exarchos
'New Orleans'	M. Exarchos	Acoustic guitar, fretless electric bass, (New Orleans) Foley, percussion, synthesisers (Korg MS20, Roli), ukulele, upright piano	M. Exarchos
'Nu Drub'	M. Exarchos	Electric bass, electric guitar, melodica, percussion, Roli synthesiser, upright piano	M. Exarchos
'Oh Buxom Betty'	M. Exarchos	Electric bass, electric guitar, synthesisers (Korg MS20, Make Noise 0-Coast, Roli), upright piano, voice	M. Exarchos
		Drums	P. Thompson
'Old Steppers'	M. Exarchos	Electric bass, electric guitar, melodica, Nord Hammond organ, percussion, Roli synthesiser, upright piano	M. Exarchos
'Oneother'	M. Exarchos, A. Zak	Synthesisers (Korg MS20, Roli), upright piano	M. Exarchos
		Electric guitar	T. Jackson
		Trumpet	M. Kirschenmann
		Electric bass	E. Santos
		Voice	A. Zak
'Outmosphere'	M. Exarchos, A. Zak	Backing vocals	N. André
		Synthesisers (Make Noise 0-Coast, Moog)	M. Exarchos
		Trumpet	M. Kirschenmann
		Voice	A. Zak
'Outta Sight'	M. Exarchos, P. Thompson	Electric bass, electric guitar, Roli synthesiser, upright piano	M. Exarchos
		Drums	P. Thompson
'Padded'	M. Exarchos, P. Thompson, R. Toulson	Synthesisers (Korg MS20, Moog, Roli), upright piano	M. Exarchos
		Drums	P. Thompson
'Psychodelic'	M. Exarchos	Drums, electric bass, electric guitar, Nord Vox Continental organ	M. Exarchos
'Ready To Chop'	M. Exarchos, P. Thompson, R. Toulson	Korg RK100S2 keytar synthesiser, upright piano	M. Exarchos
		Drums	P. Thompson

'Rebluezin'	M. Exarchos, J. Martinez, S. McGuinness	Drums, electric bass, electric guitar, harmonica, harpsichord, upright piano, voice	M. Exarchos
		Tenor saxophone	C. Hirst
		Trombone	V. Msimang
		Trumpet	D. Wilhelm
'Reggae Rock'	M. Exarchos	Drums, electric bass, electric guitar, Nord Hammond organ, percussion, Rhodes electric piano, voice	M. Exarchos
'Respin'	M. Exarchos	Harpsichord, Roli synthesiser, voice	M. Exarchos
		Drums	P. Thompson
'Rijeka'	M. Exarchos	Foley, pipe organ, voice	M. Exarchos
		Drums	P. Thompson
'Skinwalkers'	A. Caldecott, M. Exarchos	Harpsichord, voice	A. Caldecott
		Baglama, electric bass, electric guitar, Moog synthesiser, upright piano	M. Exarchos
		Flute	J. Lord
'Spin Eet'	M. Exarchos	Harpsichord, upright piano, voice	M. Exarchos
		Drums	P. Thompson
'Steppers Dub'	M. Exarchos	Electric bass, electric guitar, melodica, Nord clavinet, Nord Hammond organ, percussion, Rhodes electric piano, upright piano	M. Exarchos
'Studio A'	M. Exarchos	Drums, electric bass, grand piano, Nord harpsichord, Rhodes electric piano, voice	M. Exarchos
'Sub Conscious'	M. Exarchos	Voice	A. Caldecott
		Fretless electric bass, Rhodes electric piano, upright piano, voice	M. Exarchos
'Take 3'	M. Exarchos	Drums, electric guitar, fretless electric bass, grand piano, Roli synthesiser	M. Exarchos
'Tense Minor'	M. Exarchos	Drums, electric bass, grand piano, Korg MS20 synthesiser, Nord Hammond organ, Rhodes electric piano, voice	M. Exarchos
'To Rhodes'	M. Exarchos, P. Thompson	Electric guitar, fretless electric bass, Nord Hammond organ, Rhodes electric piano	M. Exarchos
		Drums	P. Thompson
'Toro'	M. Exarchos	Acoustic guitar, banjo, fretless electric bass, mandolin, percussion, ukulele, upright piano	M. Exarchos
'Touch'	M. Exarchos	Drums, electric bass, electric guitar, Hammond C3 organ, voice	M. Exarchos
		Drums	P. Thompson
'Train Brain'	M. Exarchos, A. Tsoukatos	Roli synthesiser, voice, Yamaha electric piano	M. Exarchos
		Electric guitar	A. Tsoukatos

(Continued)

Beats *(in alphabetical order)*	*Written by*	*Instruments sampled*	*Performed by*
'Until'	M. Exarchos, P. Thompson	Electric bass, electric guitar, Rhodes electric piano, synthesisers (Korg MS20, Moog)	M. Exarchos
		Drums	P. Thompson
'Whatever'	M. Exarchos, P. Thompson	Electric bass, electric guitar, Nord Vox Continental organ, Rhodes electric piano, synthesisers (Korg MS20, Moog)	M. Exarchos
		Drums	P. Thompson
'Wishbone'	M. Exarchos, J. Lord, M. Lord	Electric bass, electric guitars, synthesisers (Korg MS20, Roli)	M. Exarchos
		Voice	J. Lord
		Barkin' vocals	M. Lord
		Drums	P. Thompson

Produced by Stereo Mike.

INDEX

Note: **Bold** page numbers refer to tables, *italic* page numbers refer to figures and page numbers followed by "n" refer to end notes.

For Product Safety Concerns and Information please contact our EU representative GPSR@taylorandfrancis.com Taylor & Francis Verlag GmbH, Kaufingerstraße 24, 80331 München, Germany